MODEL A FORD
RESTORATION
HANDBOOK

ANNOUNCEMENT

It is with great pleasure that we are able to publish the FORD MODEL A RESTORATION HANDBOOK. I have been a Model A fan for over 35 years and have published many books on this interesting car. However, this is the first volume which has ever been written explicitly for the beginning restorer and leading him through the steps of refurbishing a Model A. For that reason it is certain to be of valuable assistance to a great many enthusiasts. The author, Gordon Hopper, has done an excellent job of describing the steps-by-step process and has omitted no detail concerning the problems as well as the methods. It is this sort of advice which makes a restoration job much easier and more pleasant.

The Model A well deserves to be preserved. Its appearance was as clean and crisp as many of the more-revered Classics, and the essential good taste possessed by Edsel Ford is clearly reflected in the whole car. It was Edsel, Henry Ford's son, who was principally responsible for the Model A, having finally convinced his father that the day of the Model T was past, and he participated fully in its design. Edsel Ford was tremendously interested in automotive styling and was the moving spirit behind the handsome Lincolns of the 'Twenties and 'Thirties as well as the famous pre-War Continental. So, there is much more behind the Model A than the average owner realized . . . none of which should take away from the fact that this was one of the most practical, inexpensive and long-lived pieces of transportation ever built.

So, if you are contemplating a Model A, or are in the process of restoration, I hope that this Handbook will help you do the kind of job the car deserves.

Floyd Clymer

April, 1966

MODEL A FORD RESTORATION HANDBOOK

Step-by-Step Procedures,
Authentic Photos, and
Clear Specifications

Gordon E. Hopper

E P B M
ECHO POINT BOOKS & MEDIA, LLC

Model A Ford Restoration Handbook was originally published in 1966, making the prices listed in this book no longer applicable. However, one can get a sense of prices in 2014 dollars by multiplying any price you see by 7.34, according to the Bureau of Labor statistics. After 2014, you can find out the multiplier rate by going to: http://www.bls.gov/data/inflation_calculator.htm.

Published by Echo Point Books & Media
Brattleboro, Vermont
www.EchoPointBooks.com

ISBN: 978-1-62654-028-6

Cover design by Rachel Boothby Gualco,
Echo Point Books & Media

Editorial and proofreading assistance by Christine Schultz,
Echo Point Books & Media

CONTENTS

INTRODUCTION

The writing of this book was undertaken during the time of a Model A restoration project and was started after problems were encountered and after the realization that actual restoration work was not covered in any of the books possessed. Floyd Clymer in his excellent book, "How to Restore the Model A Ford," has supplied the reader with a wealth of information and this book is intended to be used in conjunction with his to give a complete coverage on the subject not covered elsewhere in a single publication. It is hoped that the problems encountered by restorers will be simplified by the information contained in this book and it is written to assist the Model A Ford car owner in restoring his automobile as a do-it-yourself project and should appeal to the individual who wants the car restored with a minimum financial outlay.

If the procedure steps in this book are followed, the end result will be the possession of a SAFE automobile which should be capable of passing the necessary requirements of any state's Motor Vehicle Registry inspection and of providing many thousands of miles of enjoyable travel or transportation.

This is not a technical manual for the Model A Ford, but if reader interest in the subject goes as far as to want to know more about the theories of the actual operation of the power plant, ignition, steering, carburetion, transmission, and differential, obtain a copy of "Ford Model A Service Manual and Owner's Handbook of Repair and Maintenance" written by Victor W. Page.* This book was written in 1929 and has been revised so that it covers practically all of the Model A cars. It contains a gold mine of technical information and is well illustrated.

It is strongly recommended that the contents of this book be read in its entirety before starting on the restoration project. The author's car is a 1931 DeLuxe Coupe and naturally there will be variations between it and the other models, but the basics are the same and this book should simplify the solutions of the problems encountered in restoring most of the closed models.
The illustrations on pages 151-179 show typical models manufactured during the four year period of Model A production. The Model A Ford Album**shows all models and will help the restorer in recognizing many details necessary to an authentic restoration job.

*Clymer Publications, $7.00
**Clymer Publications, $3.50

CHAPTER 2

OBTAINING THE CAR AND THINGS TO KNOW BEFORE STARTING THE ACTUAL RESTORATION WORK

2.1 Locating the car.

Old Model A's are not exactly easy to come by with the exception of those which have already been restored and these cost between $800 and $1,000 and in some cases, even more.

However, they are available in restorable condition, but it may require a concentrated search effort to locate one. A few years ago, a lot of people had the impression that old cars were easily obtained from farmers who had stored one away in a barn or shed after purchasing a new one, probably second-hand without a trade-in. This may have been actually true at the time, but not so true today, at least in the northeastern part of the country, as the majority of them have been located and bought. At one time, automobile dealers in the northeastern section of the country scoured the area buying up all the Model A's that were available, and then shipping them to outlets in several of the southern states and to South America. Quite a few of these cars still exist in North and South Carolina, Georgia, Alabama, Mississippi, Louisiana, Texas and Arkansas.

Hot rodders on the West Coast long ago scoured the Mountain and Southwestern states for Model A vehicles and parts, but since they were more interested in the chassis than in engines, engines, transmissions and components can still be found west of the Mississippi. The best bet for the Model A fancier who wants to search is the regions of least sophistication, where automobiles are not admired as engineering masterpieces but regarded as transportation or a luxury. The backwoods, in other words, offer the best possibility for discovery of a restorable piece.

If a prospective Model A restorer or owner wishes to acquire a Model A without the long drawn-out experiences and problems of locating one, he can contact Page's Model A Garage at Haverhill, New Hampshire. This unique firm has many Model A Fords available in both restored and unrestored condition. They are in the position of not only supplying the cars, but also of being completely and professionally capable of Model A restoration and can also supply restorers with many parts.

If the search is for an engine only, a farmer might be located who has a "Doodle Bug" not being used. These were Model A chassis used as a substitute for a tractor to pull hay trailers or general farm utility work. Some of them are composite, like having a Model A engine, a 4-speed transmission, and a Model T rear end assembly. A few years ago on some farms, Model A engines were used to power saw rigs and a lot of these are not being used any longer. They can be rebuilt, if necessary, providing that the block is not cracked. Some original engines may have worn main bearings, but can be located for a reasonable price.

2.2 What to look for before purchasing.

Before purchasing a Model A, it should be looked over to some extent. If possible, fill the cooling system with water and examine the engine block and radiator for leaks. Determine whether or not the frame has not been broken and welded. Try to identfy the transmission and rear end as being Model A and not of some other manufacture. Examine the left side of the engine block for the engine serial number. These numbers are stamped into the block just above the cylinder inlet connection and the year of engine manufacture can be determined as follows:

TABLE 2-1
YEARS OF ENGINE MANUFACTURE

Serial Numbers	Year of Manufacture
1 through 5275	1927
5276 through 810122	1928
810123 through 2742695	1929
2742696 through 4237500	1930
4237501 through 4830806	1931
4830807 through 4849340	1932

At the same time, examine the block further for a plate which could have been attached by an "authorized" Ford engine rebuilder. These plates give the oversized dimensions of the pistons, the oversized dimensions of the main bearings, and the undersized dimensions of the crankshaft journals. Cylinder walls can be rebored until a specific limit is reached. Parts catalogs list pistons oversized to at least .060 inches. At this point, it is useless to continue further and the block should be replaced. However, it may be permissible to enlarge the cylinder walls and insert new wall liners. If a problem of this kind exists, it would be advantageous to talk the problem over with a reputable engine rebuilder. Crankshafts present the same problems and must be replaced when they need regrinding after being ground to something like .030 inches undersize.

A quick, easy way to tell if cylinder walls have been rebored is to remove the cylinder head and examine the top surfaces of the pistons looking for oversized dimensions which could be stamped into the flat top surface of each piston. These oversizes usually are .020, .030, .040, .060, or .080 inches.

Another check is to turn the engine over by hand, making sure that it is not frozen. If it cannot be turned over, it is not necessarily worthless, but it will require considerable work to overcome this condition.

2.3 Cost justifications.

After locating the car, the purchase price should be dictated by the condition of the vehicle. The prices will vary, of course, and it probably will be up to the discretion of the purchaser as to how many cars he will look at before buying one. But, being in short supply, one will not be able to shop around too much. If the car is in running condition and currently registered, it will command the highest cost. On the other hand, if it is partially dismantled, especially the transmission or the rear end, the price should be low. If it is determined that the engine will have to be rebuilt, it can be assumed that it will cost approximately $200 to have it rebuilt, or approximately $75 for parts (exclusive of the block) if you do it yourself. If the fenders are dented or have rotted out sections where they attach to the body, to the running boards, or to the fender braces, these will cost money to repair, or to replace, and should be considered. Examine the condition of the roof, especially if it has been repaired, including the inside wooden bows if it is a closed model. These roofs may cost $20-$30 for materials only if you do it yourself. A new roadster top costs about $40 for materials only if you do it yourself. Running boards, if they must be replaced, cost between $30 and $40 for the pair. Windshield frames, if needed, cost between $24 and $32. Original horns are very scarce and cost as much as $25. A used rear end assembly and transmission may cost $40 or $50. Cost of replacements may be greatly reduced by purchasing a parts car.

2.4 Spare parts car.

After selecting the car to be restored, if it needs fender or running board replacements, or if other major components are missing or have to be replaced, it would be advantageous to buy a spare parts car. This usually is cheaper than buying the necessary parts separately. If it is determined that one will be needed, it should be located for a reasonable price and the parts left over at the completion of the restoration can be disposed of.

2.5 **Transporting it home or to workshop.**

There are several ways of getting the car home or to where the work will be performed, depending on distance and condition of the car.

1. Register it and drive it home.
2. Have a wrecker raise the front end and tow it.
3. Pay a used car dealer to bring it home.
4. Have a salvage operator winch it into his truck.
5. Rent a low-bed trailer which you can tow behind your car.

It is strongly recommended that the car not be towed home behind another car unless it is in good mechanical condition and properly registered and insured.

2.6 **Sources of available information.**

There is a considerable amount of literature available which will assist in recognizing parts, obtaining prices, and locating items that will be needed. If there is an antique automobile restoration club in the area, help can be obtained from some of the club members. If the restoration is to be accomplished primarily by one person, information to aid will be found in the following books:

"Model A Ford Care and Maintenance"
Clymer Publications, $2.00

"Ford Model A Service Manual and Owner's Handbook
of Repair and Maintenance," Clymer Publications, $7.00

"How to Restore the Model A Ford"
Clymer Publications, $3.00

"Ford Model A Album"
Clymer Publications, $3.50

"Henry's Fabulous Model A"
Clymer Publications, $4.00

There are other informative books on the market, but these have been found to be well worth the investment.

In addition, many parts suppliers have catalogs, some of which are illustrated and show pictures of parts not illustrated to any large extent in the books above.

These firms offer their catalogs in many automobile fan magazines. Get them all, they will help a lot in identifying parts.

2.7 **Cautions on purchasing parts.**

Most parts suppliers are reputable in their business dealings.

The author has bought parts from many companies with perfect satisfaction. However, there are restorers who were not satisfied in their dealings with some companies. It was found that certain items advertised as "replacement" were a far cry from being authentic. Be careful of items advertised as "close to original," it is better to obtain them from a parts car, from a junked car, or from another restorer, in the event that the original parts are not listed anywhere.

2.8 Importance of complete dismantle.

It is possible that a restorer's intentions may be to clean the sections of the frame that can be reached, to rebuild the engine, to repair the sections of the body that are rotted out, and to paint the car, nothing more. In other words — a partial restoration. However, as a restorer becomes more familiar with the condition of the car, it should become apparent that even though the car can become operational again, it would not be restored to the best of one's ability, and it probably will also become apparent that many small items may be missing or need replacement. If this is the case, the restorer should attempt to approach as nearly a complete job as is possible within his financial means. This leads to the complete disassembly of the engine, the fenders, the radiator, the bumpers, the running boards, the wiring system, splash aprons, and brake system. Instead of removing the body completely away from the frame, it can be jacked up one side at a time to clean and paint the frame in this area and to install new body webbing between the body and the frame. It is best to do this because when the repairs or replacements are completed, the satisfaction of a job well done will be realized, and a quieter and safer car will be the result of having done a complete job. This separation of the body from the frame is explained later.

CHAPTER 3

DISASSEMBLY OF THE AUTOMOBILE

3.1 Removal of rusty hardware.

This subject may sound simple, but the removal of bolts, nuts, and rivets may turn out to be one of the most difficult parts of the entire operation. This should not be so. It will be found that old nuts and bolts continually exposed to the elements and to the highway grime have formed a strong affectation for each other that they very strongly resist separation. Bathing nuts and bolts throughout the frame and body areas with penetrating types of oil can turn out to be a wasted effort. Actually, 90 percent of the hardware can be removed with a hammer and chisel (if rusted) or occasionally, the bolts will snap off in two pieces. However, there are at least two other ways of removing rusted hardware. If there is access to a torch similar to the ones used by body shops or junk yards, this can effectively be used to heat the nut and bolt to a point where they can be separated by hand tools. Another way is to locate a specialized tool called a nut splitter. Attach it to a frozen nut, and then as it is forcefully tightened, a cutting edge will be forced into the nut, eventually cutting its way completely through. Because the nut has been forced open somewhat by the cutting edge, it usually is loose enough to be turned off the bolt or it may fall off by itself. If the bolt remains frozen to the part through which it is mounted, a large center punch or a drift pin can be held against the end of the bolt and striking it sharply with a hammer will free it. In some cases it is not too difficult to split the nut with a hammer and chisel, allowing it to either fall off or turn off. If round head bolts turn as the nut is being turned, the only solution to their removal is to cut the nut. When laying under the car removing hardware or parts, be sure to wear a pair of safety glasses. They are cheaper than paying an eye doctor to remove a piece of steel rust from an eye.

The procedures explained above generally apply to the removal of body, frame, fender, running board, bumper, and shock absorber hardware. Engine hardware at one time or another may have had oil or grease on them and should come apart by using hand tools.

3.2 Identification of removed hardware.

It is very important not only in saving time in reassembly work, but also to simplify the correct relocation of hardware, that some method of identifying all hardware as it is removed be used, even if the hardware has been ruined. Probably the simplest method of correctly cataloguing hardware as it is removed is to keep it separated into groups and identified until it is replaced. Keep them in cans, boxes, bags, or even envelopes and be sure to mark each group as to where they had been removed from. If a group of bolts resuring one particular unit turn out to be of different lengths, keep a note or sketch with the hardware telling where each piece goes. At first, this may seem unnecessary, but in the long run it will pay off by eliminating headaches and by retaining restoration authenticity.

3.3 Safety precautions.

Safety precautions while working on an automobile are of primary importance and should be observed at all times. The matter of safety will not continually referred to throughout this book, but precautions should be observed throughout the entire operation.

1. Prevent the car from moving forward or backward by placing a block of wood or a brick in front of and in back of each wheel.

2. Never go under the car while it is being held up on jacks.

3. Use heavy pieces of timber or tripods to hold the car up after it has been raised by a jack.

4. Wear safety glasses when working under the car.

Figure 3-1 is a view of the chassis shown as an aid in locating the principal parts. The shroud shown behind the radiator was discontinued during 1928.

3.4 Removal of headlamps, horn, headlamp support cross bar, horn and headlamp wiring.

The first step towards the removal of the engine is to remove the headlamps, horn, and headlamp support cross bar preparatory to removing the radiator. The headlamp and horn wiring will be disconnected first. Remove the floor mat and the floor boards. Disconnect the ground battery strap from the battery's positive post and from the frame, placing the strap, bolt and nut in a storage area.

1. Remove the headlamp wires which are held in flexible conduits from the headlamp end by pressing in on the knurled nut and turning it to the left and then removing the wires.

2. Take out the screw securing the horn dust cap, remove the cover and pull out the wires.

3. Remove the large nut that secures the horn bracket and left hand headlamp to the headlamp support cross bar.

4. Remove the left hand headlamp and the horn.

Figure 3-1 is a view of the chassis shown as an aid in locating the principal parts. The shroud shown behind the radiator was discontinued during 1928.

5. Remove the large nut that secures the right hand head-lamp to the headlamp support cross bar.

6. Remove the right hand headlamp.

7. Place both headlamps and the horn in a storage area.

NOTE: Be sure that provisions have been made to have a sizable storage area.

8. Remove two $\frac{5}{16}$ by $\frac{3}{4}$ inch bolts and lockwashers which secure each end of the handlamp support cross bar to the front fenders and their brackets.

9. Pull the headlamp cross bar up and away from the front fenders.

10. Replace the four mounting bolts and lockwashers in the two fender brackets and place the cross bar in a storage area.

3.5 Removal of the hood, hood rods, radiator, and water hose assemblies.

1. Unhook the four hood hold-down clamps and raise the left side of the hood.

2. Open the petcock at the bottom of the radiator outlet pipe between the two lower water hoses, remove the radiator cap, and allow the water to drain out.

3. Remove the cylinder water inlet and outlet gooseneck connections by removing two nuts from the engine head and two bolts from the left side of the block.

4. Push the upper and lower hose assemblies to one side, one at a time, just enough to allow replacement of the two nuts in the head and the two bolts on the left side of the block.

5. Tag the gaskets found under the inlet and outlet connec-tions for later identification and place in a storage area.

6. Remove two $\frac{7}{32}$-inch stove bolts and lockwashers from the cowl clamp that holds one end of the hood to the cowl.

7. Remove the clamp temporarily.

8. Slide the hood towards the windshield half an inch or so.

9. Lift the hood free and place it in a storage area.

10. Reassemble the two stove bolts, lockwashers, and cowl clamp back on the cowl.

11. Remove the two radiator brace rods that secure the top of the radiator to the cowl by loosening the four brace rod end nuts slightly.

12. Lift the rods out of their slotted brackets and place them in a storage area.

13. Have a helper steady the radiator while two bolts, two springs, two nuts, two cotter pins, and two rubber cushion

pads which secure the bottom of the radiator to the frame are being removed.

14. Carefully lift the radiator and its shell out and remove the hoses, the radiator outlet pipe, and the two water connections.

NOTE: After loosening the six hose clamps, if the water hoses are frozen to the radiator, the outlet pipe, or the water connections, it will be necessary to slit the ends of the hoses with a knife or razor blade in order to free them.

15. Place the six hose clamps, the radiator, the radiator shell, the radiator outlet pipe, and the two water connections in a storage area.

3.6 Removal of the radiator splash apron.

The radiator splash apron on the 1931 models is secured near the front of the radiator between the two side frame members by four sets of hardware. These hardware sets look like four bolts attached to the frame with the apron being secured by screws to one end of each of these bolts.

1. Remove the four slotted head screws and carefully lift out the splash apron.

2. Remove the four large nuts that secure the hardware sets to the frame and take out the four bolts.

3. Reassemble the four bolts, lockwashers, nuts, and screws.

4. Place the hardware and radiator splash apron in a storage area.

NOTE: The wardware used during other years is different but the removal procedures are almost identical.

3.7 Removal of water pump, fan, and fan pulley.

1. Loosen the generator arm stud nut and push the generator toward the engine as far as it will go.

2. Lift the fan belt away from the generator and water pump pulleys.

3. Remove the four nuts that secure the water pump to the cylinder head studs and pull off the water pump and fan pulley with the fan attached.

4. Tag the water pump gasket for future identification and place in a storage area.

5. Remove the fan belt, placing it with the water pump, the fan pulley, and the fan in a storage area.

NOTE: If the fan belt cannot be removed from the lower pulley, leave it alone until the engine itself is removed.

Figure 3-2 shows a front view of the engine as an aid to locate parts.

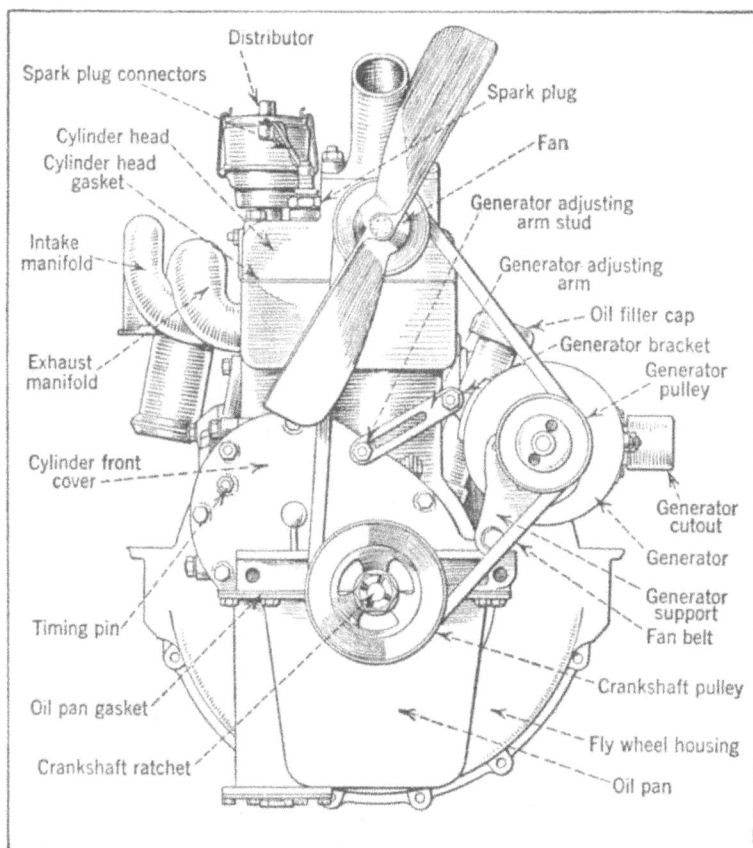

Figure 3-2. Front View of Engine for Disassembly Locations

3.8 Removal of generator and the generator cutout relay.

NOTE: Be sure that the battery ground strap has been removed as described in paragraph 3.4.

1. Remove the wires that are connected to the generator.
2. Remove the nut from the end of the generator arm mounting bolt.
3. Hold the generator in one hand and remove the generator arm mounting bolt with the other hand.
4. Lift the generator free and place it, the attached generator cutout relay, the generator arm mounting bolt and nut in a storage area.

Figure 3-3 shows a left view of the engine as an aid in locating parts.

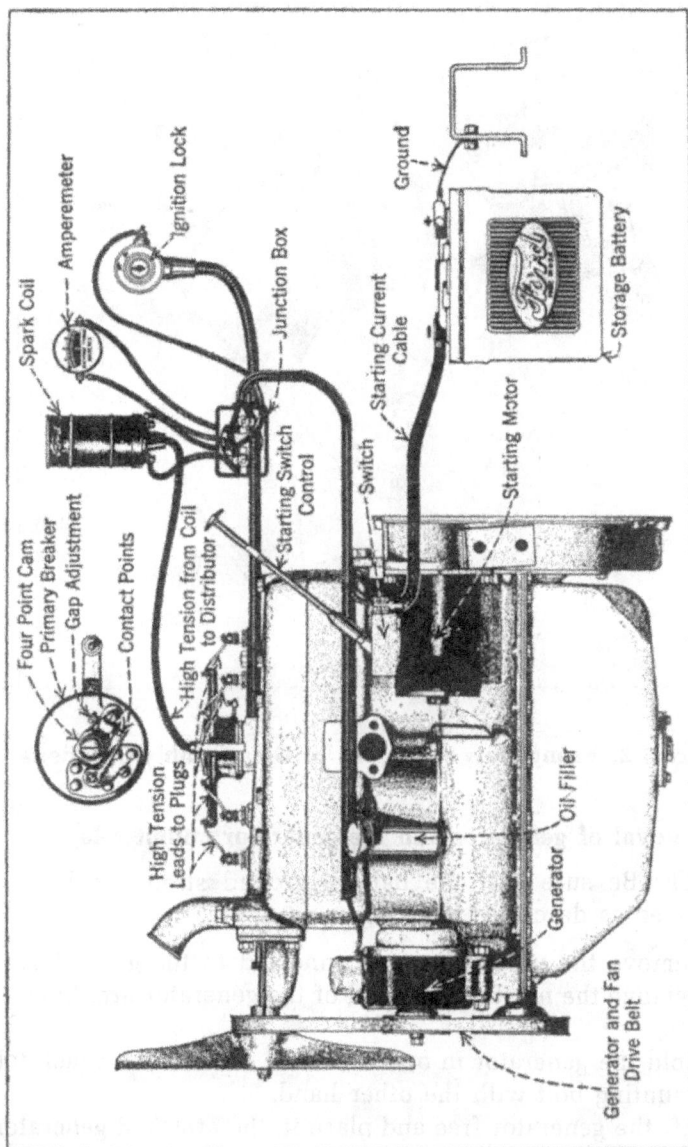

Spark Coil
Amperemeter
Ignition Lock
Four Point Cam
Primary Breaker
Gap Adjustment
Contact Points
High Tension from Coil to Distributor
High Tension Leads to Plugs
Junction Box
Starting Switch Control
Switch
Starting Current Cable
Ground
Storage Battery
Starting Motor
Generator
Oil Filler
Generator and Fan Drive Belt

Figure 3-3. Left Side View of Engine for Disassembly Locations

16

3.9 Removal of the starter motor and the starter drive.

1. Unscrew starter rod from the starter switch and remove it.

NOTE: Be sure that the battery ground strap has been removed as described in paragraph 3.4.

2. Remove the remaining battery cable from the negative battery post and the starter motor and place the cable in a storage area. If there is a battery cable clamp attached to the clutch housing, remove it and place it in a storage area.

3. Remove four screws securing the starter switch to the starter motor and place the four screws, the starter rod, and starter switch in a storage area.

4. Remove two of the three bolts that secure the starter motor to the flywheel housing.

5. Hold the starter motor with one hand and remove the remaining bolt.

6. Carefully remove the starter motor and place it with its hardware in a storage area.

3.10 Removal of gas line, linkages, choke rod, and carburetor.

1. Turn off the gas line valve if the car is equipped with one.

2. Remove the gas line from the shut off valve and the carburetor and place it in a storage area.

3. If the car is equipped with a glass filter bowl, remove it and place it in a storage area.

4. Remove the three linkages (spark control rod, throttle linkage rod, and throttle control rod) by compressing the spring-loaded ends outward and place them in a storage area.

5. Remove the spring, washer (if there is one), and coupling sleeve that secures the choke rod to the carburetor.

6. Remove the choke rod and its two rubber grommets.

7. Place the coupling components and the choke rod in a storage area.

8. Remove the two bolts that secure the carburetor to its mounting bracket on the manifold, keeping one hand on the carburetor.

NOTE: Both the Zenith and the Tillotson carburetors are removed the same way.

9. Put the carburetor and its two mounting bolts in a storage area.

10. Tag the carburetor gasket for future identification and place in a storage area.

3.11 Removal of muffler, manifold, and oil return line.

1. Remove the two bolts and two nuts from the exhaust

manifold clamp holding the muffler to the manifold.

2. Have a helper hold the free end of the muffler (or place wood blocks under it) while removing the rear tail pipe clamp from the body frame.

3. Lift the muffler clear of the car.

4. Remove the rear clamp from the tail pipe.

5. Place both sets of clamps, muffler, and the hardware in a storage area.

6. Remove the four nuts from the four studs that secure the intake manifold to the right side of the engine block.

NOTE: If the nuts are frozen to the studs, remove the nuts and their studs from the block. New nuts and studs can be used in the reassembly procedures.

7. Place the manifold, nuts, and studs in a storage area.

8. Tag the manifold gaskets for future identification and place in a storage area.

9. Remove the two cap screws that secure the oil return line to the right side of the engine below the manifold.

10. Tag the two gaskets for future identification.

11. Place the pipe and the gaskets in a storage area and replace the two bolts in the engine block.

3.12 Removal of front bumper, bumper mounting brackets, front fenders, and front fender brackets.

1. Remove the two bolts and two nuts that secure the front bumper to its mounting brackets.

2. If the bumper is secured by two oval medallions, remove two nuts, two backing plates, two medallions, and the front bumper, placing them in a storage area.

3. Remove the bolts and nuts securing the two bumper brackets to the frame.

4. Tag the brackets for future locating reassembly.

5. Place the hardware and the brackets in a storage area.

6. Remove the four hood hold-down clamps, placing them and their hardware in a storage area.

7. Remove four bolts and four nuts, securing each front fender to each running board.

8. Remove the small bolt and nut that secures the outside edge of each front fender to each front fender bracket.

9. Remove all bolts, nuts, and washers that secure the two front fenders to the car body.

10. Place the hardware and two fenders in a storage area.

11. Remove two bolts and nuts that secure each front fender bracket to the frame.

12. Place the hardware and the two fender brackets in a storage area.

3.13 Removal of throttle assembly.
1. Unscrew the foot throttle knob from its shaft.
2. Remove the two bolts at the top rear of the engine block which secure the throttle assembly to the rear of the engine block.
3. Work the foot throttle through the firewall and remove the entire assembly, placing it and the two bolts in a storage area.
4. Screw the throttle knob back on the shaft.

3.14 Removal of the wiring system.
1. Remove the four brass connector straps between the spark plugs and the distributor.
2. Remove the lead between the center of the distributor and the center of the coil.
3. Place these four straps and lead in a storage area.
4. Remove the four screws which secure the instrument panel to the cowl.
5. Disconnect the wires from the ignition switch.
6. Disconnect all wires from the ammeter.
7. Place the instrument panel and its mounting hardware in a storage area.
8. Remove all wires from the coil (on the firewall).
9. Remove the coil bracket mounting hardware from the firewall.
10. Place the coil and its hardware in a storage area.
11. Remove all wires from the terminal box.
12. Remove four bolts that secure the terminal box to the firewall.
13. Place the terminal box and its hardware in a storage area.
14. Disconnect both wires from the tail light.

15. Remove the bolts and nuts that secure the stop light switch to the frame and disconnect both wires from the stop light switch.
16. Place the stop light switch and its hardware in a storage area.
17. Unsnap the wire bail on the light switch housing.
18. Remove the light switch housing from the bottom of the steering column and pull the contact (switch) plate with the attached harness wires from the switch housing.

19. Place the light switch harness housing and bail wire in a storage area.

NOTE: The wiring harness will be replaced later with a new one.

20. Lift the distributor straight up from the engine block and place it in a storage area.

21. Remove the four spark plugs and place them in a storage area.

3.15 Removal of the engine.

There are several methods that can be used to remove a Model A engine, four of which are detailed here.

1. The most ideal method is to have an open garage area with overhead strong beams and locate the vehicle with the engine directly under one of the beams. NOTE: Chainfalls are available from a tool rental store.

2. Locate the vehicle under a large tree with the engine directly under a strong branch.

3. Build a tripod made of pieces of 4 x 4 lumber. This should be a couple of feet higher than the hood height.

4. Have an engine rebuilder remove it with a portable hoist.

5. Or, if none of the above can be accomplished, four men can lift the engine out by hand.

In any case, regardless of which method selected, it is imperative that the four wheels be blocked to prevent any movement of the car until after the engine has been removed.

1. Drain the oil from the pan, then replace the drain plug.

2. Make a check to ascertain that all accessories, linkages, and wiring have been completely removed from the engine block and the firewall, in other words, that the engine block is free and ready to be removed from the engine compartment.

3. Loop a chain with hooks on each end or a stout rope around the engine block near the front; loop another chain with hooks on each end or a stout rope around the engine block near the rear and leave them long enough to be brought together at the top center of the block.

4. Attach both of them to a hook on a chainfall which has been secured to an overhead beam, a tree limb, or a tripod placed over the engine.

5. To prevent the chains from slipping, wrap an old blanket around the engine under the chains.

6. Operate the chainfall to the point where all slack in the chains or ropes has been removed.

7. Remove the four large bolts at the rear suspensions which secure the rear end of the motor to the large brackets which

are bolted to the frame.

CAUTION: These brackets are bolted to the frame and are not to be removed from the frame until later. DO NOT remove them now.

8. Remove the two bolts at the front suspension which secure the front end of the motor to the front engine mount.

9. Remove the bolts that secure the transmission housing to the clutch housing with the exception of two widely spaced ones.

NOTE: At this point, one person should be standing by the engine and another person at the chainfall.

10. Remove the two remaining bolts and get out from underneath the car.

11. Raise the chainfal one or two notches only, just enough to lift the engine free from the areas that it rests on at the three points of suspension.

12. Pull the engine forward to clear the clutch assembly from the transmission.

13. Have a helper steady the engine while you operate the chainfall until the engine clears the frame.

14. Push the car out of the way, lower the engine to the floor or ground, remove the chainfall, and remove the chains from around the block.

15. Remove the cylinder head nuts from their studs, place the head, head gasket, and the nuts in a storage area.

16. Remove the bolts and lockwashers that secure the oil pan to the block.

17. Carefully remove the oil pan and the oil pump.

18. Place the bolts, lockwashers, oil pump, and oil pan in a storage area.

An alternative method for removing the engine would be to wrap a chain around the engine block and slip a two or three inch pipe between the chain and the block with its ends extending out over where the two front fenders used to be. After the engine mounting bolts and the clutch housing bolts have been removed, test the balance of the engine in the chain sling by raising the engine very slightly. When you have it balanced correctly, two strong helpers can lift it up and out through the front of the compartment with one or two other helpers balancing the engine to prevent it from slipping along the pipe. Be careful, and it can be done.

If the engine is removed by this alternative method, be sure to perform steps 15, 16, 17, and 18 after the engine is out.

3.16 Removal of clutch pressure plate, clutch disc, and flywheel gear.

1. Remove 12 bolts and lockwashers that secure the clutch pressure plate assembly to the housing.

2. Remove the clutch disc identifying the side that faces outward for future re-installation.

3. Place the clutch pressure plate, its hardware and the clutch disc in a storage area.

NOTE: At this point the engine should be placed on a sturdy bench for examination and/or overhaul which will be undertaken after certain work has been completed on the front end of the car. The overhaul of the engine is covered in Chapter 4.

3.17 Removal of rear fenders, running boards, running board splash aprons, and removal of body.

1. Remove the two bolts, lockwashers, and nuts that secure the end of each running board to each rear fender.

2. Remove all hardware that secures each rear fender to the body.

3. Place both rear fenders and their hardware in a storage area.

4. Remove two nuts from underneath each running board bracket.

5. Lift each running board straight up. This will draw the mounting bolts out of the brackets.

6. Place both running boards and their hardware in a storage area.

7. Remove all bolts that secure the body to the frame. (See Table 3-1).

NOTE: If the body bolts are rusted in too hard to conveniently remove, it may be necessary to use a chisel or a nut splitter to remove the nuts.

8. Identify the location of each bolt removed and place in a storage area.

9. Remove the running board splash aprons by pulling them straight out (away from the body).

10. Remove two screws that hold the steering column clamp to the dashboard. Place the screws and the clamp in a storage area.

11. Support the steering post and steering wheel in their normal position by a piece of 2 by 4 lumber resting on the ground vertically.

12. Raise one side of the body a couple of inches, either by

a chain fall or two jacks, remove the old body to frame webbing and insert a new length of replacement webbing.

13. Reinstall one running board splash apron (see paragraph 3.19), and lower the body.

NOTE: Read paragraph 3.18 before proceeding.

14. Insert new body bolts on this side of the car, but do not tighten them yet.

15. Raise the other side of the body a couple of inches either by a chain fall or two jacks, remove the old body to frame webbing, insert a new length of replacement webbing, reinstall the remaining running board splash apron, and lower the body.

16. Insert new body bolts on this side of the car and tighten down all the body bolts.

TABLE 3-1
LIST OF BODY BOLTS

Quantity	Size	Location
2	7/16 x 2³/₄ (with lockwashers)	One in each rear corner of chassis.
4	7/16 x 2¹/₄ (with lockwashers)	Under front seat, 2 on each side of the body.
4	³/₈ x 2¹/₄ carriage bolts	Near hinge end of door post, 2 bolts on each side approximately in line with the gear shift lever.
2	³/₈ x 2¹/₄ (with lockwashers)	One on each front corner of the body.
4	⁵/₁₆ x ³/₄ (with lockwashers)	Rear bumper side arm brackets to under sill at rear of body.

17. As soon as the body bolts have been tightened down, remove the piece of 2 x 4 lumber holding the steering post, and replace the two screws and clamp removed in step 10 above.

3.18 Cleaning and painting the frame.

At this point, if it is determined that the top surfaces of the frame members that actually carry the body require any extensive work, or removal of grease, etc., a decision will be have to be made on how to accomplish this.

Naturally, one choice of solution would be to use a chain

fall and a rope sling to completely separate the body from the frame. Another solution is to raise the body three or four inches and support it above the frame on wooden blocks. Either method will allow the scraping and sanding of the frame members from the firewall to the rear end before painting.

A steam cleaner can be rented and used to remove grease and oil from the frame. the transmission, the front end, the rear end, and from the engine.

The easiest way is to raise the body three or four inches and clean the top of the frame with sandpaper and apply two coats of rust-proof black enamel to it. The rest of the frame, transmission, engine, front and rear ends can be cleaned with a grease remover, Lestoil, rags, and a wire brush.

If the restorer is a perfectionist, he should obtain information on methods of sandblasting the frame sections before painting them. This is also the best day to restore the five wire wheels to their original condition.

After the frame has been cleaned and painted, lower the body and bolt it down as explained ·in paragraph 3.17.

Sand the four running board brackets and apply two coats of rust-proof black enamel to them.

3.19 Reinstallation of running board splash aprons.

After the splash aprons have been removed, examine them carefully for rusted out or weakened areas. If they appear to be in good condition, clean the top and inside surfaces and apply two coats of rust-proof black enamel. On the outside surfaces, clean only the end areas that were secured to the

front and rear fenders and apply two coats of rust-proof black enamel. Do not paint the outside exposed surfaces as this will be done when the entire car is painted.

If the aprons appear to be in a deteriorated condition especially where they secure to the front and rear fenders or to the running boards (1931), restoration or replacement will be necessary. If they are definitely bad, replace them. If only the ends are had, take them to a body shop, and have new sheet metal bonded into place on the rusted out sections. If they are not too bad, purchase a fiberglass body patch kit and do-it-yourself. If the bottom turned-up edge (1931) is quite bad, it is up to the body man, or ideally, replace the apron.

3.20 Removal and replacement of shock absorbers.

Check the operating condition of the four shock absorbers. If they are not rust-frozen and function correctly with an up and down motion of the body, they can be left alone. However, if they are not functioning correctly, remove the two bolts that secure each one to the frame and remove them.

Rebuilt shock absorbers can be purchased on an exchange basis from various Model A suppliers. If more comfort in riding is desirable and providing that the restorer is not a purist, replace the original Houdaille hydraulic type shock absorbers with tubular shocks. This type of installation requires the drilling of new holes in the frame and the restoration is no longer authentic. **Guide yourself accordingly.** Use new shock links and rubber bushings when replacing or retaining the original shock absorbers. It is best to use rebuilt original shock absorbers and retain the authenticity.

Figure 3-4 shows the location and parts of the old style shock absorbers and Figure 3-5 shows the differences between the old and new types of shock absorbers.

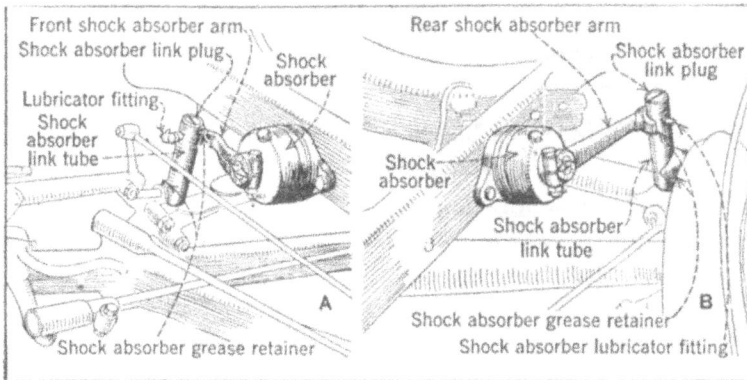

Figure 3-4. Location and Parts of Old Style Shock Absorbers

3.21 Shock absorber adjustments.

When installing shock absorbers, the pair marked CW are installed at the front right (passenger side) and rear left (driver side). The pair marked AC are installed at the front left and right rear. For the older types, the average arrow (needle valve) setting is 2 for the front pair and 3 for the rear pair. See Figure 3-5. In cold weather it may be advisable to decrease the hydraulic resistance and turn the arrow setting to 1.

On the later shocks the arrow and the numerals have been removed and adjustment is made by a slotted needle valve. During warm weather the rear valve is screwed in until it seats, then backed off 1/4 turn. The front valve is backed off

$^3/_8$ of one turn. During cold weather the rear valve is screwed in until it seats, then backed off $^1/_2$ to $^5/_8$ of one turn. The front valve is backed off $^5/_8$ to $^3/_4$ of one turn.

Resistance is increased when the needle valve is screwed in and resistance is decreased when the valve is backed out.

If the brake rod strikes the arrow type needle valve, screw out the valve and replace it with one having the newer slotted end. If it is necessary to change a valve, change all four to retain uniformity.

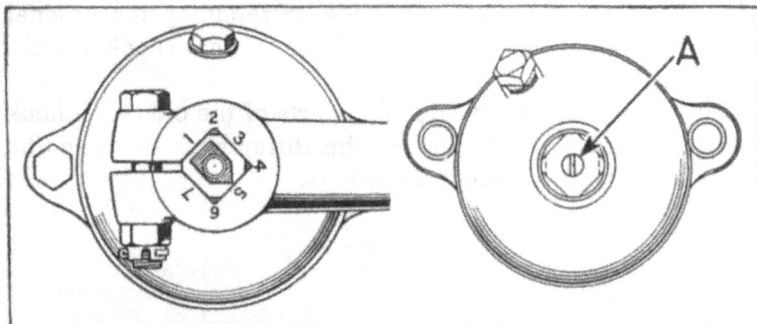

Figure 3-5. Differences Between Old (Left) and New Types of Shock Absorbers

3.22 **Removal of engine supports.**

Remove the hardware that secures the front engine support to the front cross member. Note the differences in its two flat surfaces. Place the front support and hardware in a storage area.

Remove the two brackets from the frame which supported the rear corners of the engine. Place them, their hardware, and the two rubber cushion sets in a storage area.

Loosen the bolt that clamps the pitman arm to the steering sector. Pull the arm away from the sector.

Remove the two bolts that secure the steering column to the left frame and move the column slightly away from the inside of the frame member.

Clean the remaining frame areas not done previously and apply two coats of rust-proof black enamel to them with the exception of the front cross member which will be done after certain front end work has been accomplished.

3.23 **Cleaning and painting the transmission, universal joint, torque tube, and refinishing the gear shift lever and brake handle.**

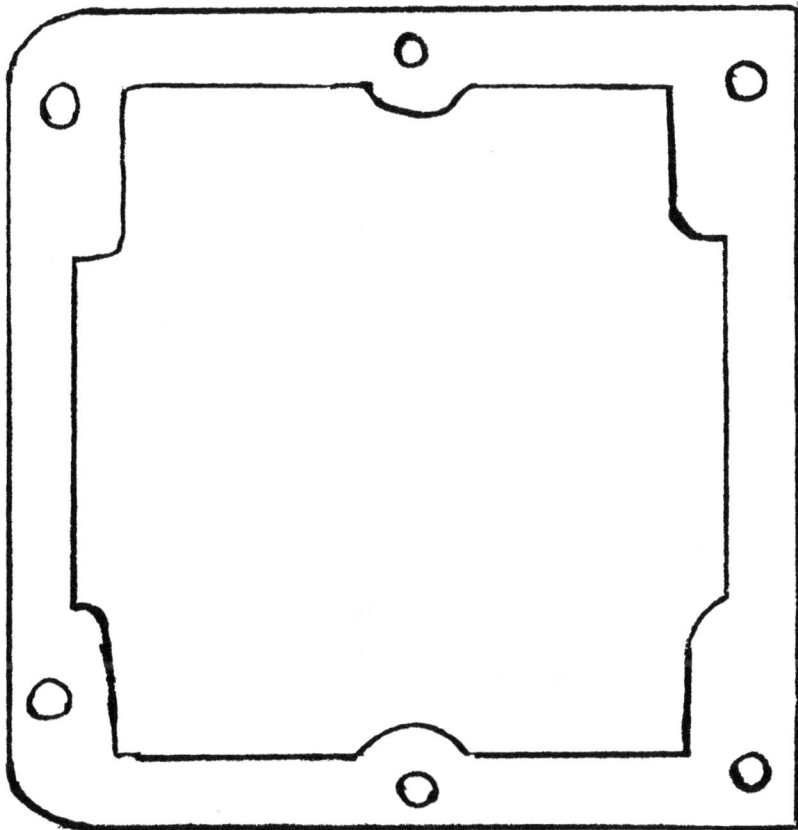

Figure 3-6. Gear Shift Lever Gasket

If the transmission appears to be OK, with no broken, worn, or pitted gears visible, now is the time to refinish its housing. Undoubtedly it will be coated with grease and dirt. Scrub it down with a grease remover, Lestoil, and a wire brush. Apply two coats of the same paint that will be used later on the engine block.

Clean the universal joint and the torque tube back to the rear end gear housing and apply two coats of rust-proof black enamel to the torque tube and two coats of engine paint to the universal joint housing.

Remove the cotter pin and clevis pin from the emergency brake handle linkage. Remove two bolts that secure the emergency brake handle to the right side of the transmission. Put the cotter pin, clevis pin, and bolts in a storage area.

Place the gear shift lever in its neutral position. Remove the bolts that secure the shift lever to the top of the transmission housing. Place the gasket, the gear shift knob, and hardware in a storage area. Figure 3-6 shows the outline of a gear shift gasket.

Take the shift lever and brake handle to a plating shop and have them rechromed (1930-1931) or renickled (1928-1929).

When you get them back, replace them carefully and use a new gasket under the shift lever plate.

3.24 Repair of the front spring and replacement of spring shackles.

1. Obtain a new set of front spring shackles.

2. If the engine is still in position, place a 2 x 4 under the oil pan and jack up the engine (after removing the four nuts on the U-bolts) until the frame is JUST starting to lift the wheels from the ground.

3. Place a short (10-12 inch) piece of 2 x 4 wood the LONG way along the axle UNDER the shackle eyes as close to the perch as possible. One on each end of the axle. Lower the jack so that the full weight of the engine and car depress the spring. This will take all tension from the shackles.

Remove the cotter pins from the shackle bolts, remove the nuts, remove the shackle back plate, and take out the shackles. It may be necessary to drive the shackles out with a drift pin.

After the shackles are out, jack the car up again and remove the wood blocks. Pull the spring down and remove it from the front frame cross member.

If the perches are bad, they should be replaced, but if they are good, drive out the shackle bushings and drive in new ones. Oil the hole in the perch before driving in the new bushings. If the old bushings won't come out, cut them in several places with a hack saw.

Saw off the rivets holding the spring clips to the spring. Clamp the spring together, placing the clamp beside the center bolt with a large C-clamp. Remove the center bolt with a drift pin. Put the spring on a bench in its normal position and loosen the C-clamp very SLOWLY.

Starting with the top spring leaf, wire brush each one, examining each one for cracks. Any cracked leaf MUST be replaced. Grind off the ridges that appear on top of each leaf.

Put a thin layer of fiber grease on top of each leaf (except the top one) and reassemble the spring. Use a new spring center bolt and clamp the leaves together. The center bolt should remain clear.

Replace the spring clips and bolt them on tightly.

Drive out the bushings in each end of the spring and replace them.

Put the two wooden blocks back on the axle and set the spring on top of them with the top of the spring just starting into the front cross member. Install one spring shackle including its two nuts, cotter pins, and grease fittings.

Leave the blocks in place and lower the engine until all the weight is on the spring. Install the second shackle, including the two nuts. cotter pins, and grease fittings. Remove the wood blocks.

Install the spring hanger bottom plate, making sure that the spring center bolt fits into the recess in the bottom plate. Install the two large U-bolts, nuts, and cotter pins.

Paint the spring two coats of rust-proof black enamel and grease the shackle fittings.

3.25 Restoration of drag link and tie rod.

1. Remove the cotter pins and threaded plugs from each end of the drag link and tie rod.

2. Remove the drag link from the front steering arm and from the pitman arm.

3. Remove the cotter pins and threaded plugs from each end of the tie rod.

4. Remove the tie rod from the two steering arms.

NOTE: If the wheels are on the car, remove the nuts securing the steering arms to the front axle and remove the drag link and tie rod as a complete assembly.

5. Remove the internal springs and washers.

6. Install two kits of link and rod end replacement parts in the opposite sequence of the removal.

7. Tighten the end plugs until they are flush with the ends. Be sure to insert the new cotter pins.

NOTE: One end of the drag link has an opening $^3/_8$-inch from one end and the other is $1^3/_8$-inches from the opposite end. Reinstall the drag link with the $^3/_8$-inch end toward the front of the car.

8. Install four new grease and metal retainers.

9. Reassemble the drag link and tie rod into their original positions and apply two coats of rust-proof black enamel to them.

10. Apply two coats of rust-proof black enamel to the pitman arm.

11. Force new grease into the arm and link through their respective grease fittings.

3.26 Restoration or repair of the steering gear.

The rebuilding steps and adjustments necessary to completely restore the steering gear are covered in a Ford Service Bulletin. It is advantageous to possess a copy of "Model A Service Bulletins" because the subject of steering is completely covered in it.

The external adjustments required by the steering gear assembly are not complicated and can be accomplished by the restorer. However, if internal parts are worn and require replacement along with a complete realignment of the steering gear, it becomes more complex and the restorer is referred to the Service Bulletin which pertains to the steering gear.

The following procedures apply to the 1930-1931 models only. Refer to the Service Bulletin for other years.

1. Remove the pitman arm from the end of the sector gear.

2. Remove all dirt and grease from the end of the sector for examination of its conditions.

3. Grasp the end of the sector and attempt to move it. If it moves up and down or back and forth, replace it and its bushings. If there is no play in the sector, turning the steering wheel from one end to the other. If there are no grinding spots or grinding bearings, proceed with step 4. If there is grinding, go to the Service Bulletin and perform a complete job.

4. Tighten the cover plate bolts (on the chassis side of the steering box).

5. Loosen the lock nut on the sector adjusting screw (on the engine side of the steering box).

6. Set the steering wheel at the center of its rotation.

7. Tighten the sector adjusting screw until it is tight, back off $1/4$ turn, and tighten the lock nut.

8. Set the steering wheel at one-half turn (either direction) from its center position.

9. Loosen the housing clamp bolt on the collar at the top of the steering box.

10. Loosen the lock nut on the bearing adjustment. (This is the uppermost nut on the steering box assembly).

11. Turn the bearing adjustment in until it is tight, back off $1/4$ turn and turn the steering wheel from one end to the other. If the steering wheel binds, back off the bearing adjustment until there is no bind in the steering wheel.

12. Tighten the housing clamp bolt and tighten the bearing adjustment lock nut.

13. Loosen the four $5/8$-inch nuts on the frame side of the steering box one-half turn each.

14. Set the steering wheel at its center position.

NOTE: Locate the cam nut at the top of the four nuts (step 13) on the chassis side of the steering box. Turning this adjustment clockwise brings the sector gear and the worm gear together.

15. Remove the play in the steering box by adjusting the cam nut and not allowing any bind in the center of the steering wheel's rotation.

NOTE: The final adjustment of the cam nut must be in a clockwise direction so as to eliminate backlash. There will always be some play in the steering gear when the steering wheel is away from its center position.

16. Tighten the four nuts which were loosened in step 13.

17. Make sure all bolts and nuts are tightened, that the steering wheel turn easily from one end to the other with no binds, and that there is no play in the center of the steering wheel rotation.

18. Reinstall the pitman arm on the end of the sector and tighten its bolt and nut.

NOTE: If the steering wheel itself is cracked, obtain a new one and replace the old.

CHAPTER 4
ENGINE OVERHAUL

The removal of the Model A engine is detailed in Chapter 3. If it has been determined that there is a major problem in the engine, then it must be repaired or rebuilt. It is a contention, that if no major engine overhaul has been performed recently, it would be a wise move to either rebuild or partially rebuild the engine at the same time that the rest of the car is being restored.

If it is decided to completely rebuild the engine, take it to a reputable motor rebuilder and have them do the work. The reason for having them do the work is because they have the automotive machine shop necessary to do a complete job including reboring the cylinder walls, grinding the crankshaft, align boring, planing the head, rebabbiting, etc. Even a good mechanic will find it is impossible to accurately perform work of this type.

If the rebuilding job is discussed with more than one rebuilder, be sure to select one who shows some enthusiasm for working on the Model A engine rather than one who seems lackadaisical about doing it, or one who tends to degrade an engine of this type.

Generally speaking, there will be at least one (and more often there are more than one) engine rebuilder in all major cities and large towns who will rebuild a Model A engine for approximately two hundred dollars.

Another approach is to obtain a rebuilt engine from one of the firms which handle them for approximately the same cost.

Here again, a local club member who has already gone down the same road can advise as to where to get the work done locally.

If it has been determined that no rebabbiting of the block is necessary, that the crankshaft is good, that the main bearing caps are good, and that the cylinder walls are true, then a partial engine rebuilding job can be done with a considerable saving of money.

Make sure that the crankcase oil has been drained, remove the oil pan, remove the oil pump, and lay the engine on its right hand side.

4.1 Removal of pistons and connecting rods.

Remove the cylinder head if it is not already off. Remove the ridge near the top of each cylinder wall with a ridge reamer tool. Remove one connecting rod cap and push the piston and rod up through and out the top of the engine block. Fasten the cap back on the end of the connecting rod and tag the assembly identifying it as either cylinder 1, 2, 3, or 4, whichever it is (cylinder 1 is the front one and they number consecutively from front to back).

Figure 4-1 locates the pistons, wrist pins, valves and valve springs.

Remove the remaining three pistons, rods, and caps the same way and tag each one.

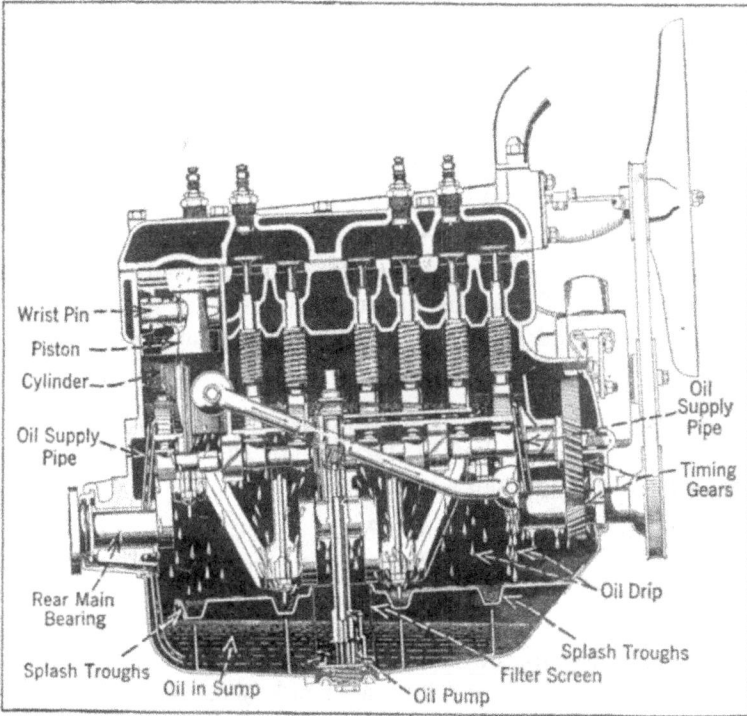

Figure 4-1. Location of Pistons, Wrist Pins, Valves, and Valve Springs

Remove the piston rings from each piston.

Examine the cylinder walls, making sure that they are not scored. Determine whether the pistons are standard or oversized. If they are oversized, the new diameter should be found stamped on top of each piston. If the cylinder walls are true, remove any glaze from the walls with a glaze breaker tool. If the cylinder walls are not true, or if they are scored, it will be necessary to have the block rebored by an automotive machine shop. Put a light coat of oil on the cylinder walls.

NOTE: The ridge reamer tool and glaze breaker tools may be rented rather than purchased.

4.2 Main bearings.

Figure 4-2 shows the location of the important structural engine parts.

Check the crankshaft for excessive play at each of the three bearing points. Check the bottom of the block at these three points and determine the condition of the babbit material. If the babbit is broken, it will have to be repoured by a rebuilder. Check the condition of the babbit in each bearing cap. If broken, the caps should be replaced and the babbit in each cap must be scraped until they fit the crankshaft perfectly with a couple of shims on each side of the cap. Scraping is explained later in the installation of the connecting rods. If the main caps appear to be in good condition (babbit material not broken and oil grooves still visible) but fit the crankshaft loosely, proceed as follows:

Place the cap in a vise between two pieces of soft wood and with a flat fine file remove a small amount of stock until a tight fit is obtained when using .003 to .005 inch shims on each end.

The caps will be correctly fitted when there is practically no drag as the crankshaft is turned. Be sure to install cotter pins at each bearing cap.

4.3 Camshaft, gears, and crankshaft.

Examine the cam shaft for excessive wear and replace it only if necessary.

Remove the timing gear cover and gasket. Examine the camshaft timing gear for broken or badly worn teeth. It should be a snug fit on the shaft and it is good policy to replace this gear with a new one.

Examine the crankshaft gear for broken or worn teeth. Replace the gear if necessary.

Install the timing gear cover and use a new gasket as shown in Figure 4-3.

Use a 1- to 2-inch micrometer and measure the four crankshaft journals. They should be 1.500 inches in diameter and they should be straight, not warped. The crankshaft is reusable if the journals are true and not more than .030 inch undersize.

4.4 Installation of pistons and wrist pins.

If the original pistons appear to be in good condition, they can be reused. If the cylinder walls have been rebored, it will

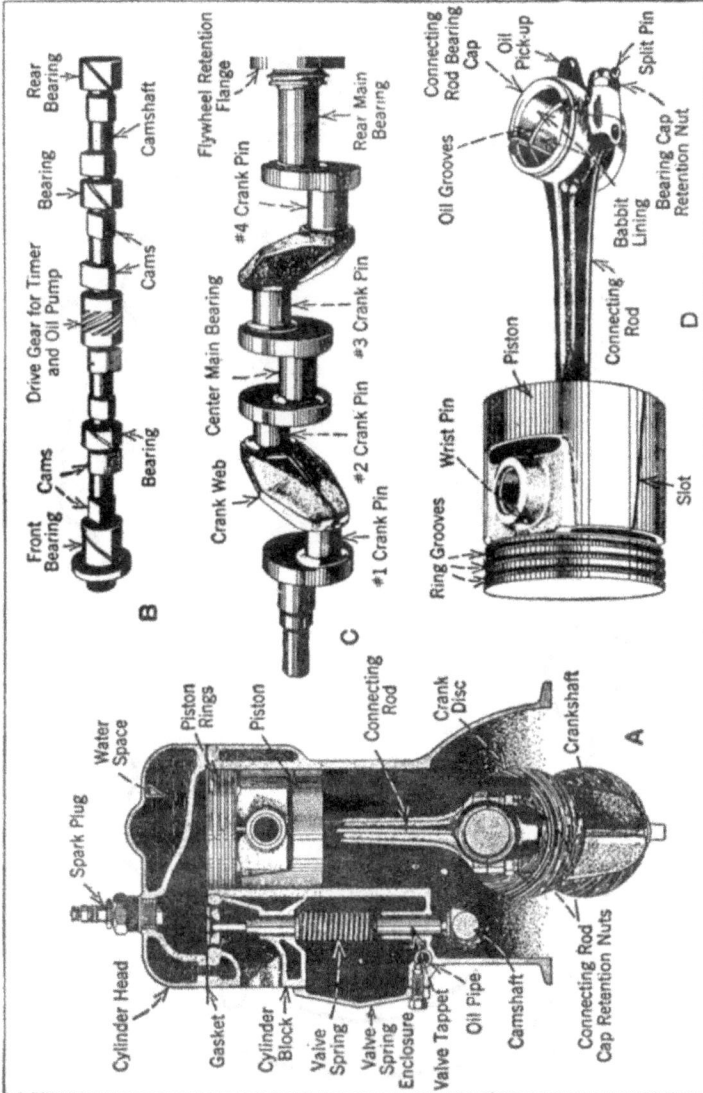

Figure 4-2. Important Engine Structural Parts

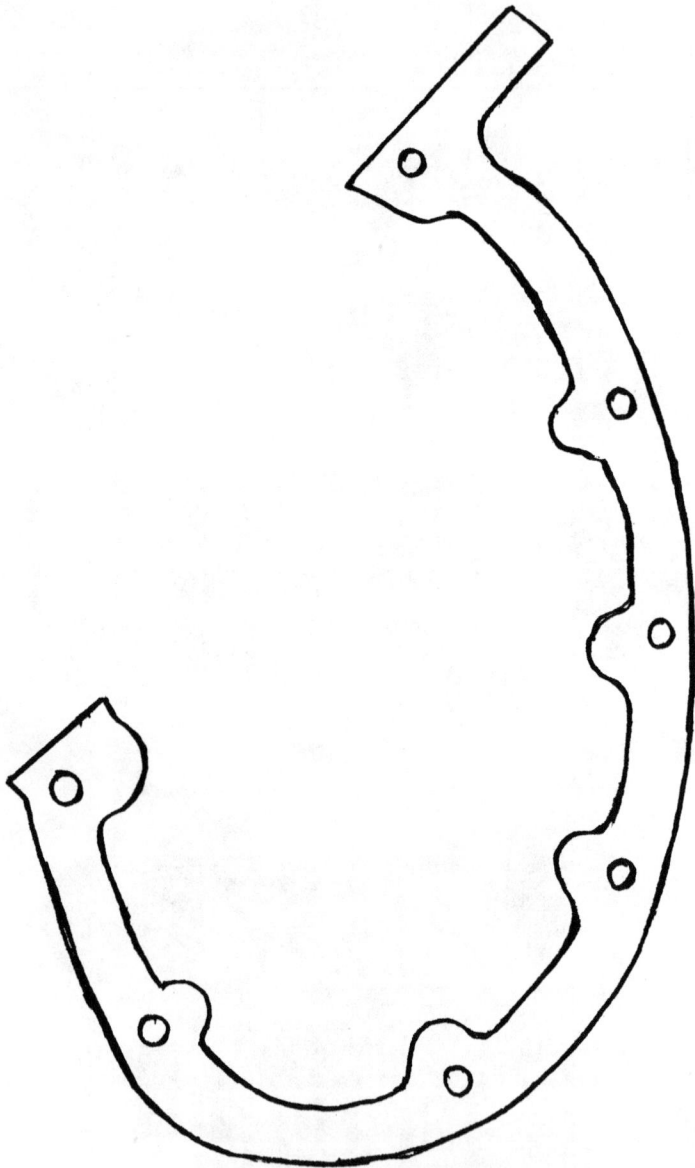

Figure 4-3. Timing Gear Cover Gasket

be necessary to use oversized pistons, of course. If the original ones are to be reused, take them with their connecting rods to an automotive shop for a bath. This will remove the carbon deposits and clean out the ring grooves including the holes in the oil ring grooves.

Obtain a new set of connecting rods and take them to the automotive machine shop with the pistons. Press out the old wrist pins with an arbor press and have the shop install the rods to the pistons using new wrist pins. It is better to have the shop install the wrist pins because they have the reamers and equipment to do a better job. However, the correct sized wrist pins can be chilled in ice water and pressed in.

If oversized pistons are used, use ones made for the amount of stock which was removed by boring, .020, .030, .040, .060, or .080-inch.

4.5 Installation of connecting rods and piston rings.

Insert one piston and connecting rod through the top of the cylinder. (The arrow on top of the piston should point forward). Cover the corresponding crankshaft journal with Prussian blue bearing compound and secure the rod to the crankshaft with its cap and .010 of shims on each end of the cap.

NOTE: The oil scoop on each cap should point toward the valve chamber.

Turn the crankshaft two or three revolutions and remove the piston. Place the rod in a vise and carefully scrape away the high spots in the babbit material indicated by the compound that has been transferred from the journal to the rod. Do the same with the rod cap. It will be necessary to follow this procedure several times until all high spots have been removed from the babbit material. A good fit will be obtained when the compound becomes distributed evenly over the entire babbit areas. You can use a knife for this job or you can grind two surfaces of a three-cornered file until a sharp edge is obtained.

It will be necessary to remove some of the shims during the scraping process and some of them should remain for future compensation. The cap should be tightened just enough to allow the rod and piston to fall by their own weight from a 90-degree position.

After all four rods have been carefully scraped in, remove the assemblies and install a complete set of new piston rings on the pistons. Purchase a set that will fit the pistons, standard .020, .030, .040, .060, or .080-inch. One set comes with steel segment oil rings for cylinders with wear, another set includes expander type oil rings for cylinders with some wear, and the third set contains plain type rings for cylinders with little or no wear. Follow the set of written instructions that come with each set of BURD rings.

To prevent breaking the somewhat brittle rings, it is best to either borrow or purchase a two band piston ring compressor. This will simplify the installation of the rings and allow all three rings to be installed simultaneously.

Apply a liberal coat of oil to the ring areas and install the pistons, rings, and connecting rods through the top of each cylinder, securing them to the crankshaft journals with the rod caps. Don't forget to use cotter pins here and remember that the oil scoops on each cap point toward the valve chamber.

After tightening each connecting rod cap, turn the engine over to make certain that there is an increased drag due to the installation of the new rings. If any piston fails to drag it is because the glaze on the cylinder walls was not broken previously.

After the car has been driven a few hundred miles, it will be necessary to remove some of the shims under each cap as the babbited areas may become worn in more evenly than was obtained during the hand scraping process.

4.6 Removal and replacement of the valves.

The valves may be ground or replaced while the engine is out of its compartment, or they may be reworked without removing the engine. If the engine is out, perform steps 12 through 25 only.

1. Drain the water from the radiator.
2. Loosen the two hood rods at the radiator ends.
3. Pull the radiator slightly forward and lift off the hood.
4. Remove the spark and throttle rods.
5. Disconnect the choke rod.
6. Shut off gas line valve and remove the gas line.
7. Remove the carburetor and the manifold.
8. Loosen the generator and remove the fan belt.
9. Disconnect the spark plug connectors and remove the distributor.
10. Remove the cylinder head complete with water pump and fan.
11. Remove the oil return line.
12. Remove the valve chamber and clean out the valve chamber with gasoline and a paint brush.
13. Compress the valve springs with a valve lifter and re-

move the valve spring seat retainers. See Figure 4-4.

14. Remove the valve springs.
15. Remove the valve guide bushings.

16. Push the valves up from their seats and examine them for grooves or pitting and examine the block for cracks in the seating areas.

If there are cracks or check marks in the block, it will have to be taken to an automotive machine shop for the installation of new inserts.

If the valves have been ground in before, or if any of them are burned, it will be necessary to replace the valves. If the block has been burned, the machine shop can resurface the seating areas. In most cases, it is best to obtain a new set of valves, valve guides, and valve springs and do a complete job.

17. Put an abrasive paste between the valve head and seat and turn the valve so that the cutting material removes the roughness from the valve seat and valve, ftting one to the other.

18. Lift the valve from its seat frequently as the grinding process continues to evenly distribute the valve compound.

19. The accuracy of the fit is found by spreading a film of Prussian blue pigment over the valve seat. Drop the valve in place and turn it about $1/8$th turn with the tool. If the seating is good, both the valve head and seat will be uniformly covered. If the surfaces are uneven, high spots will show the coloring, and the low spots will show no coloring.

10. Install the valves and insert the valve guides into the cylinder block.

21. Grind the valves again lightly to obtain perfect seating.

22. Check the clearance between the valves and the push rods. The correct clearance should be .001 to .014-inch.

23. Clean out the chamber making sure that no abrasive remains.

24. Replace the valve springs with new ones.

25. Install the valve chamber cover using a new gasket. Figure 4-5).

4.7 Radiator repair.

If there are no leaks in the radiator it can be used again and installed later. If there are leaks, or if it appears dirty or rusty, take it to a radiator shop for cleaning, repair, and testing. Check

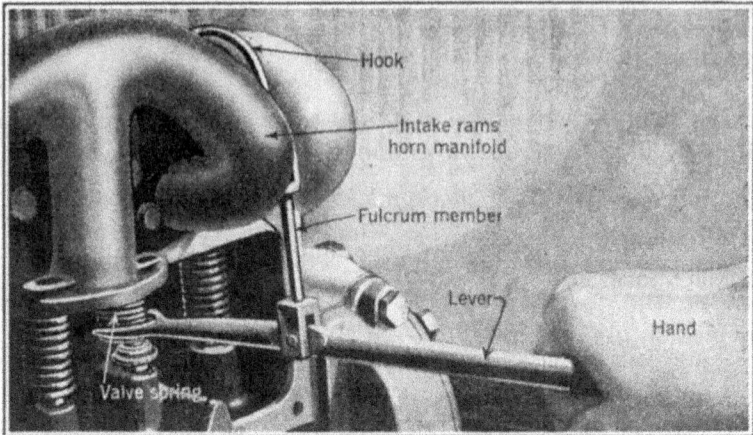

Figure 4-4. Use of Valve Spring Lifter

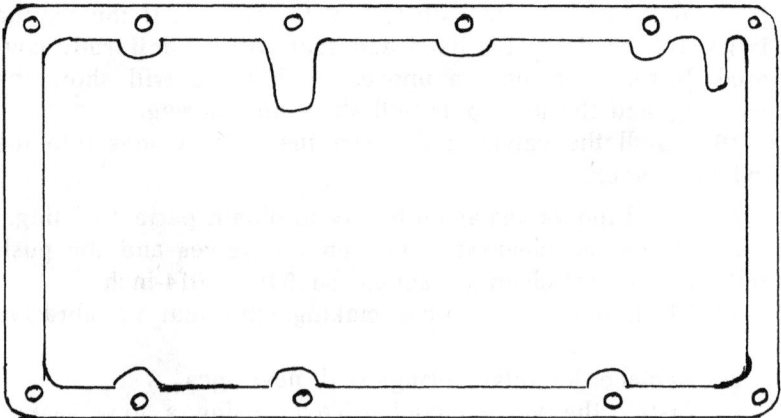

Figure 4-5. Valve Cover Plate Gasket

the four brackets that secure the radiator shell to the radiator. If they are loose, have them welded. Check the two lower brackets by which the radiator itself is secured to the frame. If they are loose, have them welded. If the overflow pipe has come loose, have it welded into place. In the event that the radiator is not reusable, new ones are available from several sources.

4.8 Flywheel gear checks.

Examine the ring gear on the flywheel very carefully. Model A engines usually stop in two or three positions on this gear and the teeth in these areas may be bad. If they are worn or broken, obtain a new ring gear. Take the flywheel to a black-smith shop or automotive machine shop, have it heated, and remove the ring gear. Heat it up again and install the new ring gear. Install a new flywheel housing gasket.

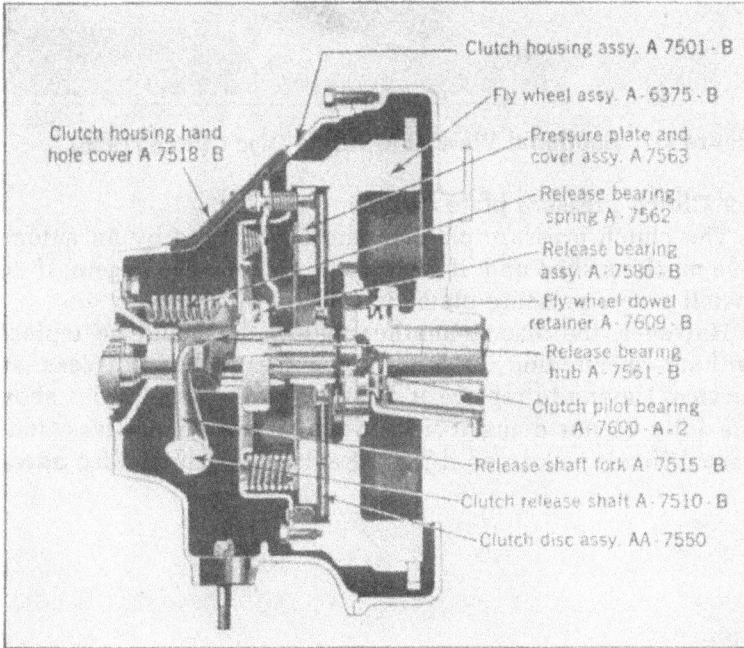

Figure 4-6. Construction of Newer Type Clutch

CAUTION: Do not install the engine until after the work on the transmission has been completed.

Early 1928 models used a multiple disc type of clutch (see Figure 4-7).

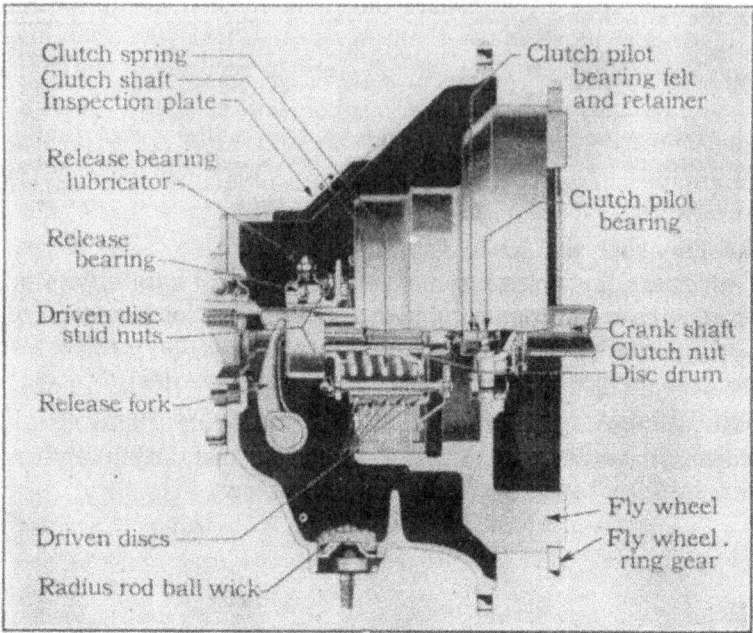

Clutch spring
Clutch shaft
Inspection plate

Release bearing
lubricator

Release
bearing

Driven disc
stud nuts

Release fork

Driven discs

Radius rod ball wick

Clutch pilot
bearing felt
and retainer

Clutch pilot
bearing

Crank shaft
Clutch nut
Disc drum

Fly wheel
Fly wheel
ring gear

Figure 4-7. Sectional View of Multiple Disc Clutch (1928)

4.9 Clutch pressure plate and disc.

The clutch pressure plate should be checked by an automotive machine shop and, if it is good, it can be used again. If not, install a new pressure plate. See Figure 4-6.

However, the disc is another matter. It should be replaced without questioning its condition unless it has not seen any service. Having to replace it after a few thousand miles shows the folly of not doing it now when it is a simple task and a very difficult task to do it later. Replace the clutch disc **now.**

CHAPTER 5

RESTORATION OF THE TRANMISSION
AND DIFFERENTIAL

5.1 General discussion.

It is imperative that the condition of the transmission be determined before reinstalling the engine. Should steps not be taken to correct any apparent defects in the transmission system, it is possible to have much trouble later on. Even though the Model A transmission was a good system at the time when it was designed and even though the gears, bearings, and other parts in the transmission are of good quality material and workmanship, it **is** possible that any or all of the gears, shafts, or bearings can be pitted, broken, or excessively worn due to the carelessness of previous owners of the car.

If the restoration is continued without checking the transmission, then, at a later date troubles may be determined and it will be necessary to remove the engine again or that the rear end of the car will have to be removed to correct the transmission problems.

The flywheel and clutch assemblies were removed with the engine, so this provides an access into the open end of the transmission housing.

1. Remove the six bolts that hold the gear shift handle and plate to the top of the transmission gear box. Remove the gear shift and plate.

2. Open the plug at the bottom of the gear box and drain out the lubricant. Flush the box out thoroughly with gasoline.

3. Examine the condition of all the gears in the gearbox very carefully. If any of them have broken or missing teeth, or if any of them show excessive wear of teeth, then they should be replaced. If they are not worn or broken, then it is conceivable that the transmission has already been replaced.

5.2 Removal of the transmission.

1. Place wooden blocking under the transmission housing (to prevent it from dropping to the ground after it becomes separated from the transmission) and under the transmission box to relieve any strain on it later.

2. Remove the emergency brake handle.

3. Remove the service brake rod from the brake pedal.

4. Remove the radius rod support and separate the front radius rod from the universal joint.

5. Remove the lower end of the speedometer cable from the end of the torque tube near the universal joint.

6. Looking into the open end of the transmission housing, remove the four bolts that secure the transmission housing to the transmission case.

7. Lift out the transmission housing and both foot pedals as a complete assembly.

8. Remove the blocking that was under the transmission housing.

9. Remove the housing that is around the universal joint exposing the inner housing and the universal joint.

10. Remove the gear box and the universal joint as a complete assembly by pulling it towards the front of the car. The universal joint will slide off the rear main shaft.

11. Remove the bolt securing the universal to the end of the transmission main shaft and pull off the universal joint.

12. Remove the inner housing, replace the gasket between the housing and the gear box, and reinstall the inner housing if the original gear box is being reused, or to the new gear box.

Determine whether more than one or two of the gears will require replacement. Determine whether the shafts and bearings will require replacement. If no more than two gears require replacement, disassemble the gears from the box, and replace the bad ones. See Figure 5-1 for location of gears in the box.

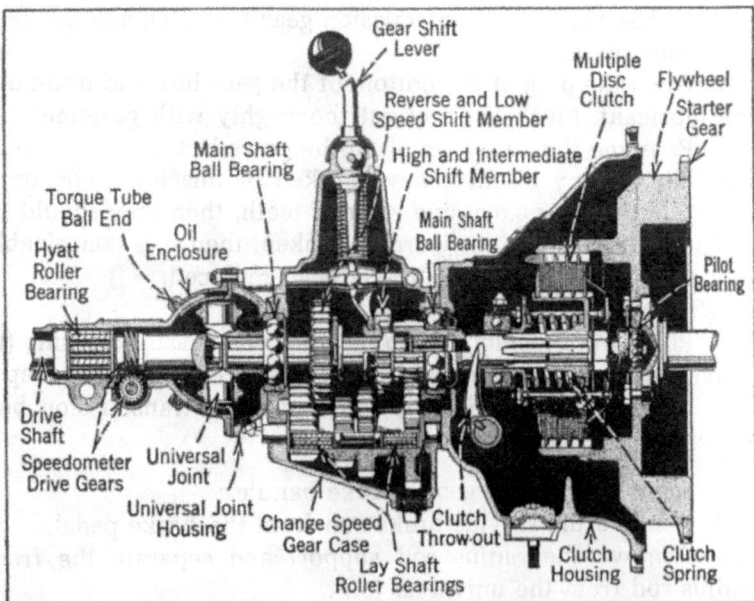

Figure 5-1. Sectional View of Transmission

If, however, more than two gears require replacement, obtain a rebuilt transmission, and replace the entire unit.

5.3 **Installation of the transmission.**

Grasp the free end of the drive shaft that goes to the differential. Turn it and observe the amount of free play. (You may have to consult a Ford mechanic who can tell how much play is permissible). If the amount of play is acceptable and if there are no strange noises coming from the differential, it can be assumed that no major work is required in the differential and work may be continued in the transmission. If trouble is apparent in the differential, stop the work at the transmission and correct the problems in the differential.

1. Bolt the universal inner housing to the gear box and put a new gasket between them. Run safety wire between the four bolt heads.

2. Grease the spline on the transmission main shaft and push the shortest splined end of the universal over the shaft.

3. Install the bolt that secures the front of the universal joint to the transmission main shaft.

4. Lift the gear box to its location under the car.

5. Grease the spline on the end of the rear main shaft and fit the longest splined end of the universal joint over the shaft and push the gear box towards the rear of the car.

6. Put blocking under the gear box.

7. Line up the holes in the universal joint inner housing with the holes in the two sections of the outer housing and bolt them together.

8. Bolt the transmission housing and pedals to the front of the transmission gear box.

9. Tighten the drain plug under the box and fill the box to the required level with 600W lubricant.

10. Position the sliding gears and the two forks so that the gearshift handle and plate can be reinstalled in a neutral position. Install the gearshift and plate.

11. Reinstall the front radius rods, using a new support kit.

12. Connect the service brake rod back to the brake pedal with its clevis pin and a new cotter pin.

13. Reconnect the speedometer cable.

14. Secure the emergency brake handle to the side of the transmission and connect its brake rod using a clevis pin and a new cotter pin.

Move the gearshift handle around, making sure that it will drop correctly into its four positions. If not, you will have to re-position the sliding gears.

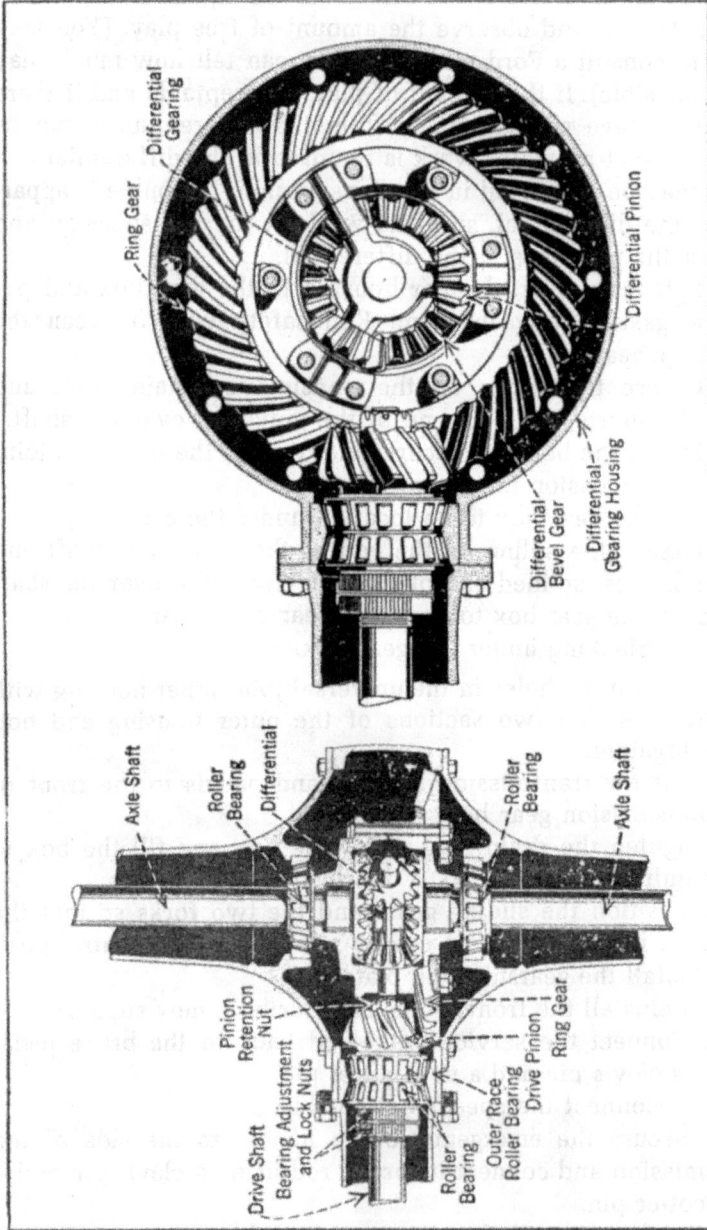

Figure 5-2. Sectional Views of Bevel Driving and Differential Rear Axle Gearing

Ring Gear

Differential Gearing

Differential Pinion

Differential Bevel Gear

Differential—Gearing Housing

Axle Shaft

Roller Bearing

Differential

Roller Bearing

Axle Shaft

Pinion Retention Nut

Drive Shaft Bearing Adjustment and Lock Nuts

Roller Bearing

Outer Race Roller Bearing

Drive Pinion

Ring Gear

5.4 **Differential.**

If the differential has been kept correctly lubricated throughout the years, the rear end should require no work. However if there are grinding sounds in the rear axle or differential, or if the rear wheel action is sloppy, it will then be necessary to facilitate repairs.

1. Refer to a copy of the Model A Service Manual and Owners' Handbook.

2. Remove all brake rods going to the rear wheels.

3. Remove the shock absorber links.

4. Remove the bolts that secure the speedometer gears to the drive shaft. Remove the gears.

5. Remove the universal joint housing.

6. Remove the two U-bolts that secure the rear spring to the frame.

7. Jack the car up high enough to allow removal of the rear axle and torque tube by supporting the front end of the torque tube and rolling the whole assembly out from under the car.

An explanation of the disassembly, repair, and reassembly of the entire differential system is thoroughly covered in the **Model A Service Manual and Owners' Handbook, Chapter XII.**

Figure 5-2 shows the bevel driving and differential rear axle gearing in cutaway views.

CHAPTER 6

INSTALLATION OF THE ENGINE

6.1 **Replacement of front and rear motor mounts.**

Before installing the motor, obtain a new front motor mount kit and a complete set of rear motor mounts. Install the front mount on the front frame cross member as shown in Figure 6-1.

Remove the two brackets on the side frames which support the rear corners of the engine. Install new rear rubber mounts. The two square pieces go between the outside of the frame and the outer square metal pieces.

Apply two coats of Ford Green (Model A) engine paint to the rear wall of the engine block.

Apply two coats of rust-proof black enamel to the engine side of the firewall.

6.2 **Installation of the engine.**

Raise the engine and position the car so that the engine can be lowered into its compartment.

Lower the engine very carefully into the engine compartment. The rear of the engine must be fitted into the transmission before it can be correctly placed on its three suspension points.

Bolt the two front corners of the engine to the front mount and bolt the two rear corners to the two brackets that are supporting the rear of the engine. Use cotter pins.

Remove all rigging used in lifting and lowering the engine.

Install the bolts and lockwashers that secure the flywheel housing to the transmission.

Install the foot throttle assembly. This is tricky, but it can be worked into place and bolted to the rear of the engine.

Install the cylinder head using a new head gasket (Figure 6-2) and draw the head nuts down evenly. Plug up (or mask over) the water inlet and outlet holes, the water pump hole, the distributor hole, the carburetor hole, the oil return line holes, and the spark plug holes and apply two coats of Model A Ford Green engine paint. If the engine is painted with a brush, it may not be necessary to protect the holes.

6.3 **Muffler installation.**

Obtain a new muffler (tapered type) and paint it with two coats of a 600 degree silicone enamel paint. Apply two coats of heat resistant exhaust manifold paint to the manifold. This paint will withstand 1000 degrees of temperature.

Install the muffler and the manifold, using a new manifold

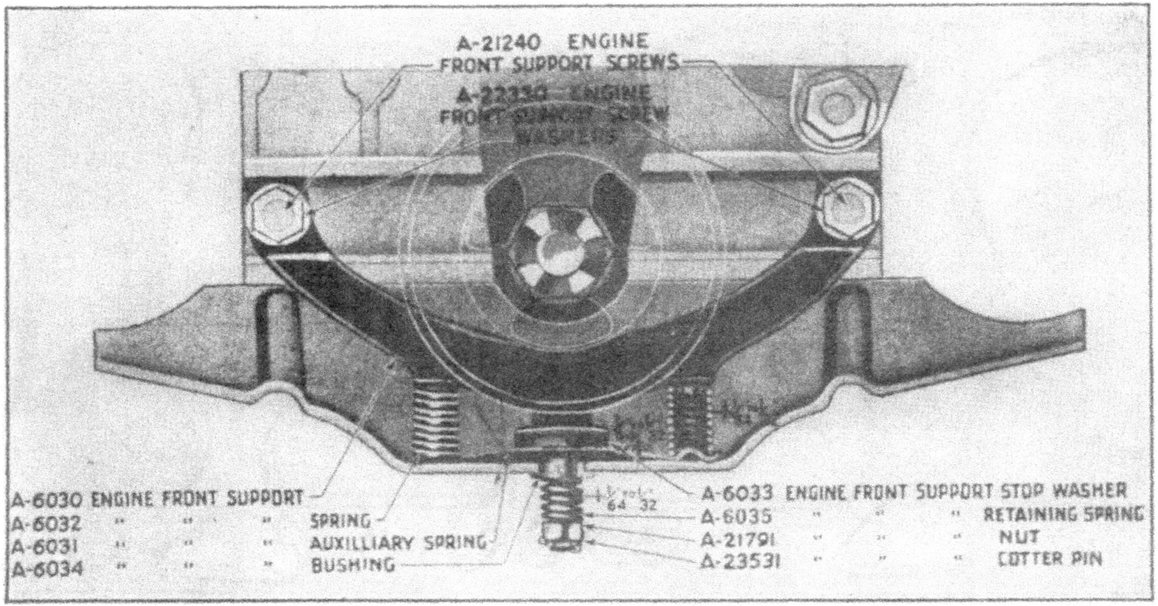

Figure 6-1. Installation of Front Motor Mount

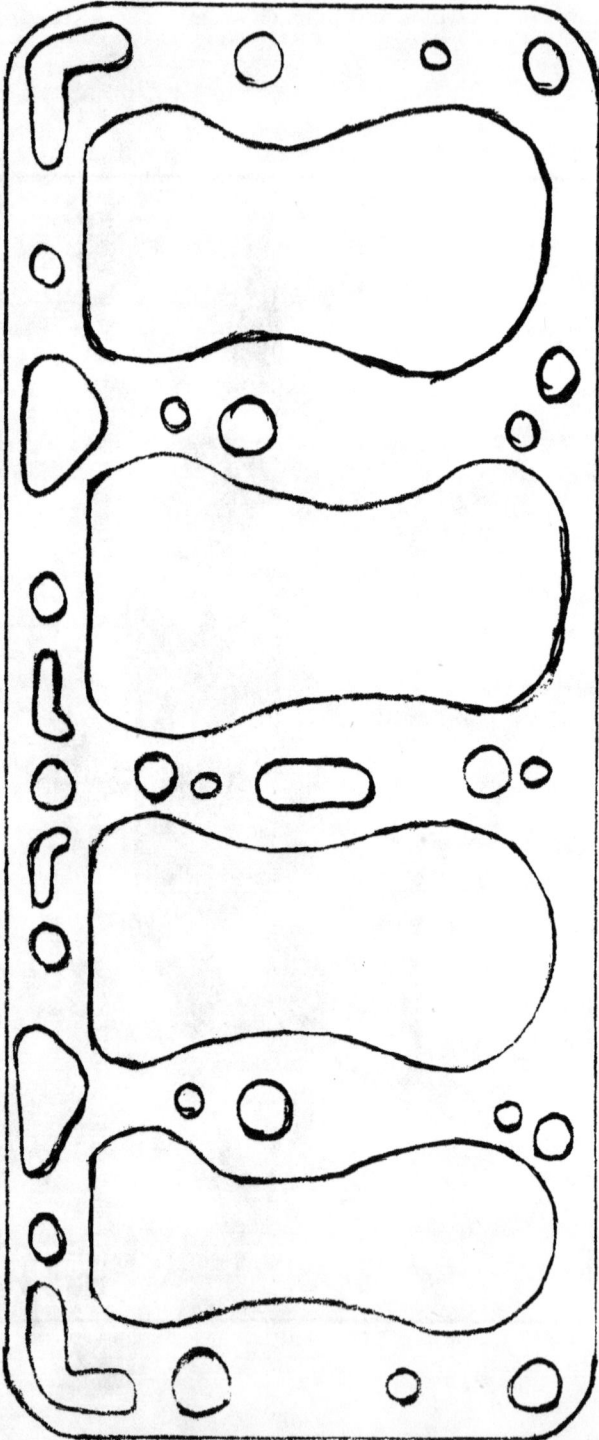

Figure 6-2. Cylinder Head Gasket

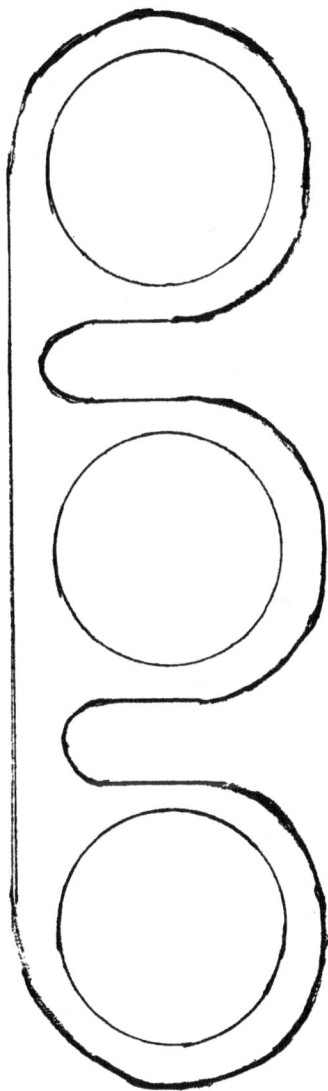

Figure 6-3. Manifold Gasket

gasket (Figure 6-3). Paint both the muffler clamps with the M250 paint.

6.4 Installation of oil pump and oil return line.

Apply two coats of rust-proof black enamel to the oil return line and install it on the right side of the engine block, using two new round gaskets (or washers).

Disassemble the oil pump. Clean it well, including the screen. Replace any worn parts that are evident. Reassemble it and replace the gasket (Figure 6-4). Install it up through the bottom of the block. If it does not stay in place, it will then be necessary to install it simultaneously with the oil pan.

Figure 6-4. Oil Pump Gasket

6.5 Installation of oil pan.

Clean the oil pan thoroughly and apply two coats of rustproof black enamel to its outside surfaces. Install new gaskets at both front and back of pan where it has been cut to fit against the crankshaft. Install the oil pan (and the oil pump) using new oil pan gaskets (Figure 6-5).

6.6 Carburetor installation and restoration.

Install the carburetor, using a new gasket between it and the engine block. If it is a Zenith carburetor, obtain a rebuilding kit and replace the internal parts. If it is a Tillotson carburetor, open it up and make sure that the float is working correctly. Make a new internal gasket, using the original as a pattern. If necessary, a replacement Tillotson carburetor can be obtained.

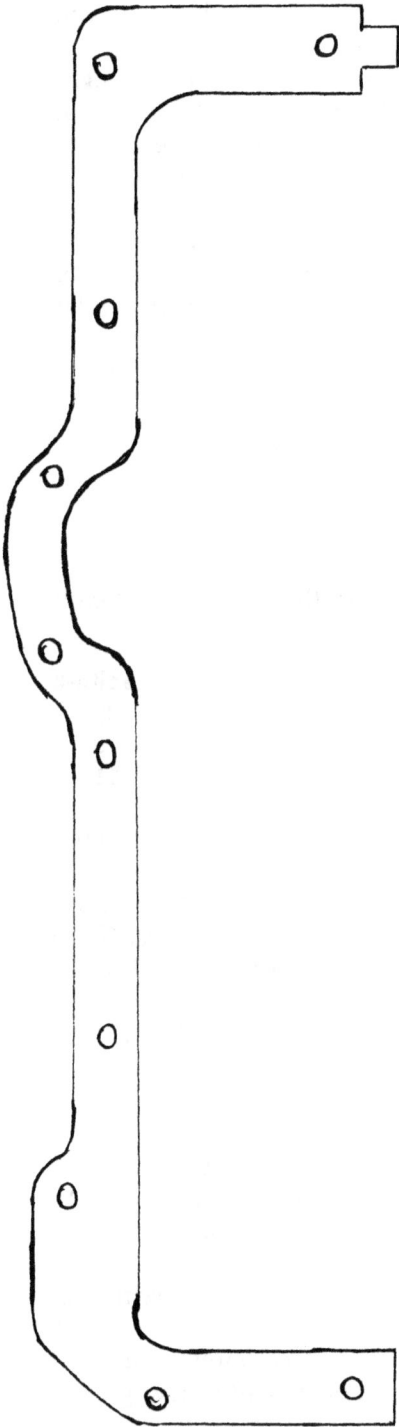

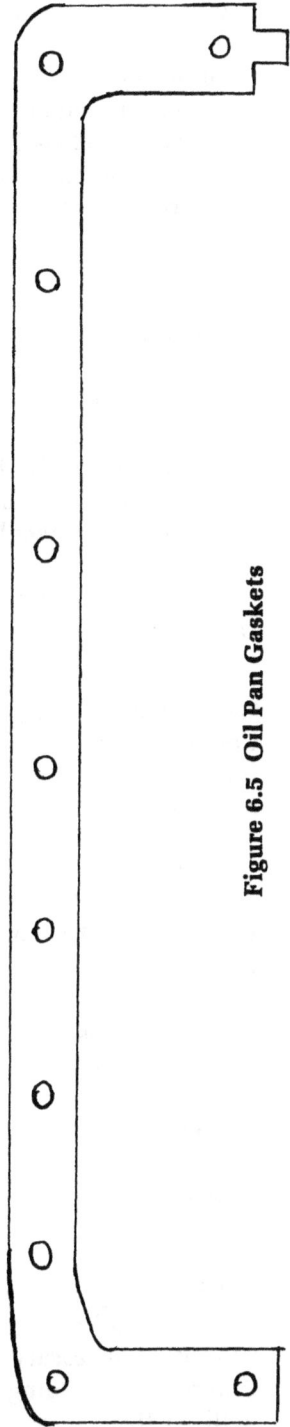

Figure 6.5 Oil Pan Gaskets

It will operate more efficiently than the Zenith but it is **not** authentic. Guide yourself accordingly.

Three types of carburetors were used in Model A's throughout the four years of production and they were built by Zenith. The first model had a double venturi, no name on the outside of the casting, and was used for the first eight months in 1928. The next model used a single venturi and carried the stamp "Zenith-1" on the bowl. In early 1930, the idle needle seat was made integral and the bowl carried the stamp "Zenith-2." In mid-1931, the filter screen was eliminated and the fuel inlet was moved to a position under the idle screw.

The most common problems that exist in the carburetor are:
1. Missing, or leaky gaskets.
2. Incorrect float level.
3. Dirt in the jets.
4. Worn needle valve.
5. Worn needle valve seat.
6. Sticky or dirty float valve.

Individual Ford part numbers in the three types of Zenith carburetors are as follows:

	Double Venturi	Zenith-1	Zenith-2
Cap (secondary jet)	19	21	20
Compensator	18	19	19
Idle jet	10	11	11
Main jet	20	$19^{1}/_{2}$	20
Throttle plate	20	$18^{1}/_{2}$	$18^{1}/_{2}$
Venturi	24 (or 10)	843	843

Remove the lower bowl section and the venturi. Place everything **except the venturi** in a cleaning bath for 2-3 days. If the original jets are not worn excessively, they can be used in the restoration of the carburetor. If they are worn or damaged, it will be necessary to obtain a rebuilding kit. The jets should last a lifetime and should not require replacement unless they have been abused.

The main jet is the one standing upright in the lower bowl section. The cap jet is the one on an angle besidet he main jet. The idle jet is the long one (on an angle) coming from the upper casting. The compensator is the one in the base of the float chamber having a screwdriver slot in one end.

Obtain a gasket kit and replace all the gaskets, in particular, the main and cap jet gaskets.

If it is necessary to rebuild the entire carburetor, obtain a complete rebuilding kit and a new flot. Usually, new gaskets will suffice.

To prevent wearing down the needle valve and its seat in the future, adjust the flot level to a position where the engine runs normal (when warm) with the choke open $1/4$-turn and with an additional washer under the shut-off valve so it will be lowered. This will eliminate the continual necessity of screwing down the choke rod.

Carburetor jet parts authenticity is determined by the fact that the original parts were numbered, replacement parts are not.

Make sure that air holes and passages are clear before re-installing the carburetor.

The float level is adjusted by bending the arms of the float or its bracket.

6.7 Installation of gas line, choke rod spring and sleeve, distributor, spark plugs, linkages, crankshaft pulley, and water pump.

Obtain and install a new gas line between the carburetor and the gasoline shut-off valve on the firewall (1931). Actually, a glass filter bowl mounts on the firewall of 1928-1930 and the gas line connects the filter bowl to the carburetor. In the 1931 models, the glass filter bowl mounts on the carburetor and the gas line connects the shut-off valve to the filter bowl.

Obtain and install a new choke rod spring and sleeve set.

Install the distributor after replacing any obviously bad parts. Use new points, condenser. and rotor regardless of their appearance.

Install new spark plugs and four new spark plug copper connector straps.

Install the three spark and throttle linkages.

Install the crankshaft pulley and ratchet, water pump, and new pump gasket (Figure 6-6).

Install the generator after installing new brushes in it. If the generator shaft appears worn or chewed up, replace the generator with a new or rebuilt unit.

6.8 Installation of generator.

Position the generator loosely and close to the block and install a new fan belt. If it uses a two-piece crankshaft pulley, there may not be enough clearance between the edges of the pulley and the front frame cross member to insert the fan belt. If this is the case, loosen the front motor mount bolts and jack up the engine until enough clearance is obtained. After the belt is in place, pull the generator away from the engine block until the belt slack has been removed. Tighten the generator.

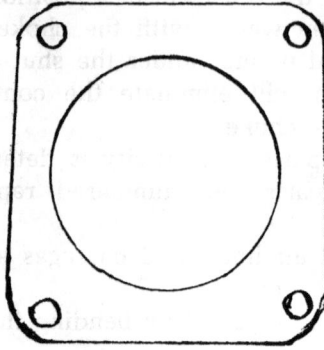

Figure 6-6. Water Pump Gasket

6.9 Installation of miscellaneous parts.

Install the fan blade on the water pump shaft and paint it two coats of black rust-proof enamel.

Apply two coat of black rust-proof enamel to both the inlet and outlet water castings, use new gaskets (Figure 6-7), and install the two castings.

Paint the water pump and generator two coats of rust-proof black enamel if necessary.

Replace the brushes in the starter motor, paint it black, and install it.

Paint the oil filler pipe two coats of rust-proof black enamel.

6.10 Engine splash pan installation.

Paint the engine splash pans two coats of rust-proof black enamel and attach them to the bottom of the block with some of the oil pan bolts and to the bottom of the side frame members.

6.11 Radiator installation.

Attach the radiator shell to the radiator using four new bolts, nuts, and lockwashers. Secure the radiator loosely to the frame using a new set of hardware, rubber pads, springs, and cotter pins.

Attach the two hood rods to the top of the cowl and fasten them loosely to the top of the radiator.

Obtain a radiator shell lacing kit and replace the old shell lacing.

Obtain a cowl lacing kit and replace the old cowl lacing.

Install the engine hood in its respective brackets on top of the cowl and on top of the radiator shell.

If the hood fits well, tighten the radiator mounting bolts, install the two cotter pins, and tighten the two hood rods on top of the radiator shell. Paint the rods two coats of rust-proof black enamel.

Make sure that there is a gasket in the radiator cap and if the cap is in good condition, install it; if not, replace it.

6.12 Installation of moto-meter (water temperature gage).

If the radiator temperature gage is missing, try locating one in a junkyard. The correct one has a diameter of slightly more than two inches and will have the Moto-meter Co. name and address on one of the internal plates and Ford on the other plate, both visible from the outside. Obtain a new set of plates and replace the old ones. The outside of the unit can be cleaned up, if necessary, on a cloth buffing wheel. Drill a hole in the radiator cap to accomodate the moto-meter and install a new gasket.

6.13 Installation of water hoses.

Either cut three new water hoses, using the old ones as patterns, or obtain a new set of correct water hoses and six new hose clamps. Install a short hose on each end of the metal section that has a petcock on it. Install this as an assembly with the petcock on the bottom and two loose clamps on each end between the bottom radiator water connection and the outlet connection on the left side of the engine block. Tighten the four hose clamps. Slide two clamps loosely onto the longer hose, install it between the top radiator water connection and the inlet gooseneck on the top front of the block. Tighten the two clamps, fill the cooling system with water, and correct any leaks which may develop.

6.14 Radiator splash pan installation.

Install a piece of $^3/_4$-inch thick wool felt material on top of the radiator splash pan where the old one was (there are two small rivet holes for this). Place the pan into its original position between the side frame members and secure it with its original hardware if they are reusable. If not, make up a new set or obtain new sets.

6.15 Timing the engine.

The spark must occur at the end of the compression stroke and the timing must be checked from that point. To time the

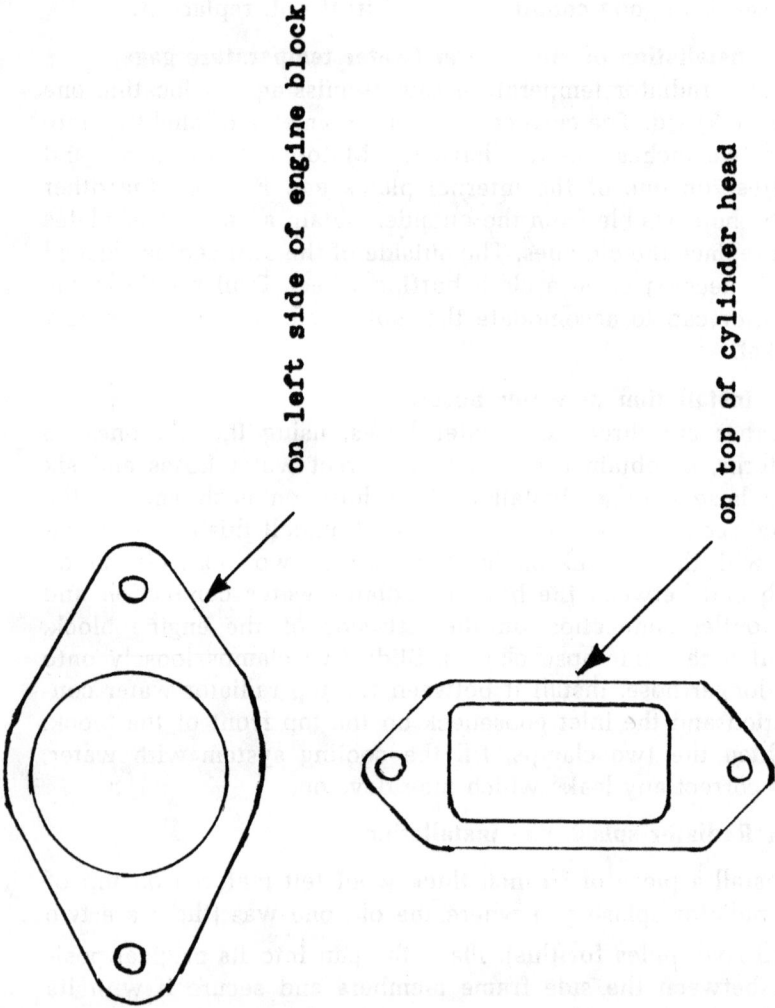

on left side of engine block

on top of cylinder head

Figure 6-7. Water Connection Gaskets

spark, it is necessary to adjust (or check) the breaker contact points and to time the ignition.

6.15.1 Setting up the breaker points.

NOTE: If the points are burned or pitted, dress them down with an oil stone or replace them. If setting up new points for the first time, use a new set of points, a new condenser, new distributor body and cover, and a new rotor arm.

1. Lift off the distributor cap, rotor, and body.

2. Turn the engine over slowly with the crank until the breaker arm (Figure 6-8) rests on one of the four high points of the time with the points fully open.

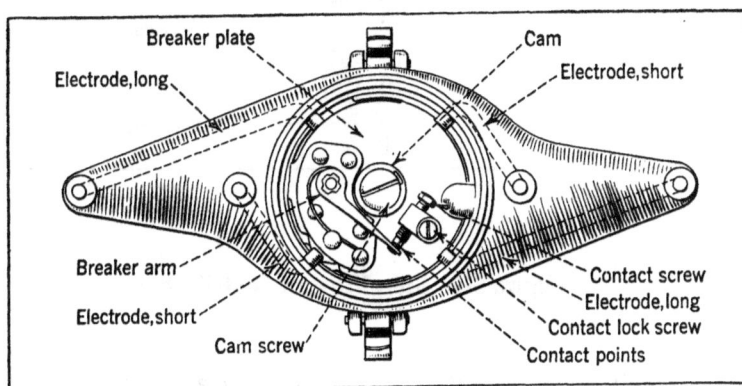

Figure 6-8. Top View of Distributor

3. Loosen the lock screw and turn the contact screw until the gap between the contact points is .015 to .018 inches.

4. Tighten the lock screw and proceed to time the ignition as follows:

6.15.2 Ignition timing.

1. Screw out the timing pin which is stored in the front of the timing gear cover and insert its opposite end into the opening.

2. Turn the engine over slowly with the crank, pressing in firmly on the timing pin at the same time.

3. At the end of the No. 1 piston's compression stroke the pin will slip into a small depression in the camshaft gear.

4. Loosen the cam locking screw on top of the cam until the cam can be turned.

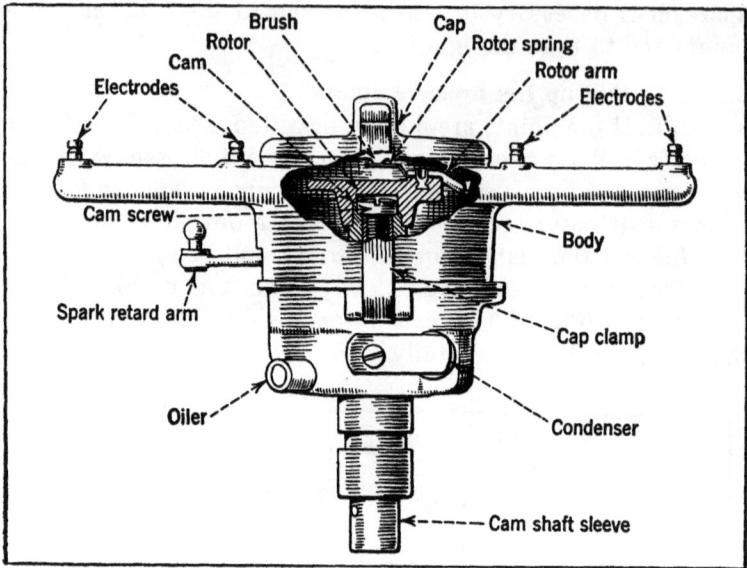

Figure 6-9. Side View of Distributor

5. Install the distributor rotor and body and turn the rotor until the rotor arm is opposite the number one cylinder contact point in the distributor body.

6. Withdraw the rotor from the cam, turn the cam slightly in a counterclockwise direction until the breaker points just start to open, and tighten the cam locking screw tightly.

7. Install the rotor and the distributor cover.

8. Remove the timing pin from the timing gear and screw it back tight into the timing gear cover.

CHAPTER 7
RUMBLE SEAT REPAIR AND RESTORATION

It is conceivable that the sheet metal and the seat cushions in the rumble seat compartment will be found in a state of disrepair or deterioration. Also, it is possible that this area may have been restored before by a previous owner.

7.1 Preliminary examination.

1. Remove both rumble seat cushions.

2. Examine the condition of the seat platform, the inner curved panel (inside rear), the floorboard, and the sill assembly that extending along both side (inside) of the body from the firewall to the rear-most member between the two side frame members (directly below the bottom edge of the rumble seat cover opening).

3. If the sill assembly is in good condition, wire brush it and apply two coats of rust-proof black enamel to its entire surface areas.

4. If the floor board is in good condition, clean it up, and apply two coats of rust-proof black enamel.

5. If the remaining interior sheet metal is in good condition, wire brush it and paint as above.

6. If the metal panels at each end of the rumble seat (at the rear wheels) are in good condition, wire brush them and paint as above.

If deterioration predominates, proceed as follows:

7.2 Restoration of the body sill.

1. Remove the seat platform and floorboard.

NOTE: Usually the sill sections extending from the back of the front seat to the rear end of the car will be rotted or rusted out, while the forward sill sections will be in good enough condition to remain in the car after being wire brushed and repainted.

2. Remove the bad sections of the sill. This can be accomplished by cutting the sill sections with a hacksaw just behind the front seat, remove the two rear body bolts, remove the bolts that secure the sill sections to the rear-most cross section, and remove the floorboard channel cross piece to the sill sections. Retain the two hardwood blocks that were at the rear pair of body bolts as they will be reused. Drill out the rivets on both sides of the vertical metal braces where they were secured to the sills.

3. Determine the necessary lengths of the new sill sections

and make allowance for a one-inch overlap on the front.

4. Have new sill sections made up by a local sheet metal shop using the removed sections as patterns.

NOTE: An alternative method would be to locate a sill in usable condition from another **identical model.** Remove the necessary lengths and after installation have the two front ends welded to the ends of the two front sections that were not removed.

5. Install the new sill sections by inserting them into the rear cross member, drill new holes using the holes in each end of the cross member as a pilot, and secure the rear of the sills with 8-32 machine screws, nuts, and lockwashers.

6. Drill new holes in the sills for the two rear body bolts.

7. Insert the two hardwood blocks and bolt the sills to the frame side members with new rear body bolts.

8. The front ends of the new sills should be welded (or bolted) to the ends of the two front sections that were not removed.

9. Position the vertical braces, drill new holes through the sills, and attach the braces using 8-32 machine screws, nuts, and lockwashers.

10. Install the floorboard channel cross piece (or replace it if necessary) and secure it with 8-32 machine screws, nuts, and lockwashers.

NOTE: If the car is a roadster, check the condition of the seat riser behind the front seat and replace it if necessary.

11. Remove grease and dirt from the new sheet metal pieces, and apply two coats of rust-proof black enamel to the new pieces, to the vertical braces, and to the channel piece.

12. Wire brush the inside surfaces of the rear-most frame cross member and apply two coats of rust-proof black enamel.

NOTE: All interior sheet metal body work and painting should now be finished with the exception of the panels beside each rear wheel.

7.3 Restoration of the rear wheel panels.

1. If the two rear wheel panels are in good condition, wire brush them and apply two coats of rust-proof black enamel.

2. If they are rotted or rusted out along their bottom surfaces they must be repaired or replaced. If the panels are not too bad, a body shop can replace the bad areas. If they are generally bad all over, they will have to be replaced.

3. Saw out the old panels, retaining about three inches along the outer contour.

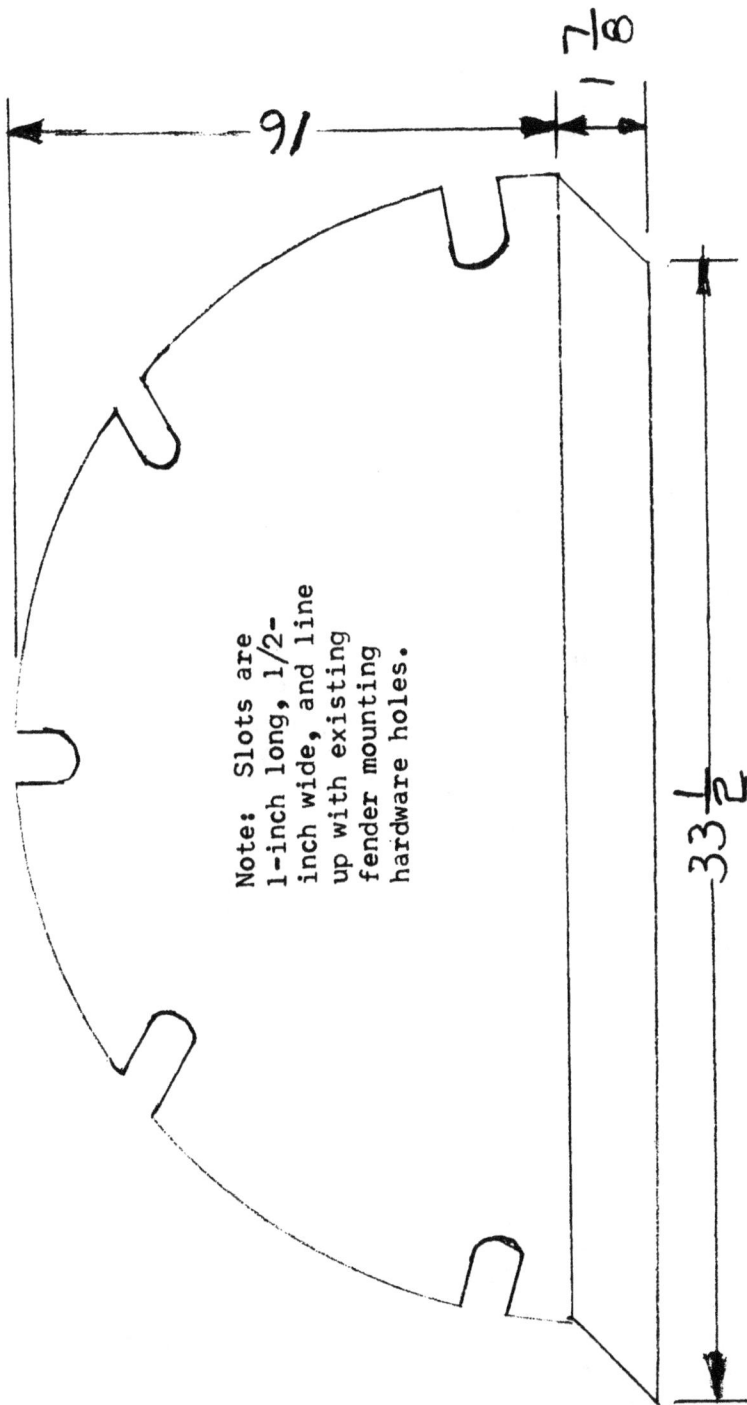

Note: Slots are 1-inch long, 1/2-inch wide, and line up with existing fender mounting hardware holes.

Figure 7-1. Panels for Under Rear Fenders

4. Obtain a new set of rear wheel replacement panels and secure them to the retained contour pieces using new fender mounting hardware. This hardware will be removed later and used to secure the two rear fenders. A more economical way of replacing these two panels would be to have them made up in a local sheet metal shop. However, the authenticity will be lost because it is doubtful whether the shop could duplicate the grooves found in the original panels. Judge yourself accordingly. If you have the panels made up locally, Figure 7-1 can be used as a pattern.

5. Apply two coats of rust-proof black enamel to both the inner and outer surfaces of the two new panels.

7.4 Inner curved panel.

1. Remove the inner curved panel at the rear of the compartment if it is rotted or rusted out.

2. Obtain a replacement panel, paint both surfaces of it, install it, and repaint the exposed surface of it, if necessary.

3. If the curved outer panel (under the rumble seat opening) has deteriorated, replace it.

7.5 Rumble seat floorboard.

1. If the floorboard is to be replaced, obtain a duplicate metal one or make one from ³/₄-inch thick marine plywood. The approximate dimensions are 19 x 38 inches. Work the edges down to obtain a good fit. Wood to metal squeaks can be eliminated by installing strips of felt material along all the edges of the floorboard.

2. Drill at least six holes in the new wooden floorboard in line with the existing holes in the sills and the channel pieces and secure the floorboard with 8-32 machine screws, flat washers, nuts and lockwashers.

3. If the metal floorboard is used, secure it with 8-32 machine screws, nuts, and lockwashers.

4. Apply two coats of black rust-proof enamel to the metal floorboard or two coats of porch and deck enamel to the wooden floorboard.

7.6 Replacement of platform section.

1. If the original platform section is reusable, secure it to the sills using self-tapping screws.

2. New platform sections are available and preferably should be used. However, the old one can be used as a pattern and a new one made up by a sheet metal shop. But, of course, it will not be authentic. Attach the new platform section to the sills

using self-tapping screws and apply two coats of black rust-proof enamel on all exposed surfaces of it.

7.7 **Hardware for the rumble seat.**

1. Examine the condition of the lower set of brackets and bumpers in the rumble seat compartment (these are the ones that prevent the cover from opening too much). If the brackets are good, clean and paint them, replace the rubber bumpers, and attach them to the platform section. If these brackets are missing, obtain a new set and install them.

2. Examine the condition of the cover hinges and brackets. If they are broken or missing, replace them.

3. Apply two coats of black rust-proof enamel to the inner surfaces of the rumble seat cover.

4. Open the rumble seat cover and apply two coats of black rust-proof enamel to the channels in the body into which the cover seats itself when closed.

5. Install two new round rubber bumpers on the body near the two upper corners of the rumble seat cover opening for the cover to close against.

7.8 **Repair or replacement of rumble seat cushions.**

If the rumble seat cushions have been restored or replaced, reinstall them. If the cushions are missing or in poor condition, a new set of foam rubber units should be obtained. If the original springs are still in reusable condition, obtain the upholstery material and do-it-yourself, or take them to a local car upholstering shop. Used springs may be located, but they are hard to find.

Install both seat cushions.

Make a lightweight cardboard pattern of the exposed sides of the inside body from the rumble seat to the back of the front seat. Cut a piece of hardboard to the contours of the pattern and upholster it. Attach these two kick panels to the inside of the compartment, using sheet metal screws and cup washers. The use of screws is superior to using clips and the panels will remain in place for many years.

It may be possible to locate new rumble seat kick panels.

Install a new rumble seat floor mat.

CHAPTER 8

REPAIR AND REASSEMBLY OF THE BODY PARTS

The removal of the fenders, running boards, lights, horn, and bumpers were discussed in Chapter 3.

8.1 Fender repairs.

If there are any large rotted out holes in the outside surfaces of the fenders, it will be necessary to have them repaired in a body shop. It is possible to patch them with fiberglass or body filler but it is not recommended. However, there are other repairs that the restorer can make to the fenders. If it is decided to use fiberglass, obtain a repair kit from an automotive supply house and follow their directions. If it is decided to use body filler, get a can of it including the dryer and follow the instructions on the can. With a careful grinding and sanding job, it is possible to obtain a satisfactory appearing fender, but it is conceivable to question the length of its usefulness due to the vibration of the car.

The best way to repair a bad fender is to have a body man cut out the rotten areas, weld or rivet in new pieces, and then grind it down evenly to match the contour of the fender.

If the edge of the fender that attaches to the car body or the edge that attaches to the running boards is bad, the restorer can repair this by one of two methods.

1. Remove the rotted or rusted areas, cut pieces of fiberglass to fit the contour of the curves or straight angle pieces to fit against the running boards and install the fiberglass.

2. Cut pieces of metal to fit the contours of the curves and have them welded underneath the fenders. The piece to fit the ends of the fenders that attach to the running boards should be slightly contoured and welded to the bottom edge of the fenders.

Drill new holes in the mounting areas to mate with the holes in the body or with the holes in the running boards.

To obtain absolute authenticity, replace the bad fenders with new ones.

8.2 Running board repairs.

If the flat surface of the running boards are bad, a body man can weld in new pieces and grind them flush with the rest of the running board. If there are many holes in the running boards, the restorer can affect a temporary repair, but to retain the authenticity, the running boards should be replaced with new ones. Complete new running boards are available. In the event that only the running board trim is to be replaced, this is also available.

If the pieces that secure the running boards to their brackets are bad or missing, other ones can be taken from other defective running boards and welded into place.

Remember that body filler or fiberglass on a running board is only a temporary solution and that replacing the running boards with new ones is the only authentic way to obtain a solution.

8.3 Body repairs.

It is practically impossible for the average layman to do any outside body work if it is to be done the way it should be done. Rotted out sections of the body will have to be repaired by a body shop man. Sections of the lower panels under the rear quarter windows (coupe) just above the running board aprons have a tendency to deteriorate more than the upper body. Pieces that are four feet long and about six inches wide with the correct contour are available from at least one of the smaller parts suppliers in Massachusetts. Two of them will be enough to repair the two sections below the quarter windows and the two lower sections in front of each door (coupe). If these areas are rotted out, obtain a pair of the replacement pieces and have the body man install them.

8.4 Repair of the running board aprons.

Again, this area must usually be repaired by a body man. If the aprons are real bad, they must be replaced. If the apron areas under the body (against the frame) are bad, fiberglass patches may work. If the outside surfaces are good but the curved piece that locks into the running board (1931) is bad, it is very difficult to repair the curved piece and the apron may have to be replaced. If the areas that attach to the fenders are bad, they may be repaired by fiberglass, necessitating the drilling of new matching holes. Fiberglass fenders and running board aprons are available on the market, but do not use them unless the metal ones cannot be located.

8.5 Installation of the running boards.

Attach both running boards to their respective brackets using a new set of running board mounting hardware.

8.6 Installation of fenders.

Install the four fender brackets.

Attach all four fenders to the body and to the frame using

fender welting between the fenders and the body and attach them to the running boards using a complete new set of fender mounting hardware which is available from several suppliers.

8.7 Reassembly of lights, horn, and bumpers.

Install headlamp support and tie rod assembly on top of the two front fenders and secure it with four new sets of mounting hardware. Bolt the right front headlight to the headlamp support and tie rod assembly. Attach the left headlight and the horn bracket to the headlamp support and tie rod assembly. Replace the patent plate and diaphragm on the horn, if necessary. (See paragraph 11.4).

Attach the front bumper to the front bumper braces using the correct oval head medallions and backing plates. All Model A's use the same medallions to secure the two outside ends of the front bumper. 1928 and 1929 A's use a round medallion ornament in the middle of the bumper, 1930 and 1931 A's use a small oval medallion in the middle of the front bumper.

Attach the rear bumper or bumperettes to the rear bumper brackets using oval head medallions identical to the ones securing the front bumper. If bumperettes are used, install the bracket (or bar) between the two rear bumper braces.

The bumper end bolts and spacers should be replaced with new ones.

Check the condition of the two rumble seat step plates and replace them after painting the car, if necessary. A gasket to go under the fender step plate is available, if necessary.

8.8 Wheels and tires.

Early 1928 models used AR type wheels which must be retained on the early models. Later 1928 and 1929 models used a 21-inch wheel with 4.5 x 21 tires. 1930 and 1931 models used a 19-inch wheel with 4.75 x 19 tires.

19-inch wheel scan readily be distinguished from 21-inch wheels as the outside edges of the 19-inch wheels resemble the outside edges of the modern-day wheels, whereas the 21-inch wheels have a heavy beading around their outer edges.

Today, the tires and tubes used on the Model A's are readily available from the Firestone Tire and Rubber Co. Original molds and specifications have been retained by Firestone, the original producer of Model A tires, and the tires are currently being manufactured. Production techniques have increased the life span of the new tires by ten times that of the original tires, they are made of high-grade rubber, they give better blowout protection, and they give longer trouble-free service. Original size tubes are made from Butyl rubber and hold air ten times longer than tubes made of natural rubber.

When new tires and tubes are required, contact a Firestone dealer. If they are not in stock, they can usually be obtained in less than two weeks.

CHAPTER 9
REPLACEMENT OF THE ROOF

9.1 General.

If the car being restored is of the open type, it undoubtedly has, or needs a fabric roof. In the event that the roof of an open car is to be replaced, there are at least two approaches to doing it.

1. Obtain a new fabric roof kit and install it yourself. Instructions and pictures are generally furnished by the manufacturer to assist in installing the roof.

2. Have it done by a professional restorer.

The roofs of many Model A's were not kept in good condition down through the years and many of them were simply ignored and allowed to deteriorate. This becomes evident to the restorer after the headlining or windshield visor have been removed. Many of the wooden side pieces or cross pieces will be found in a rotted or generally poor condition because of water leaks in the fabric roof or because of many years exposure to the weather elements. In many cases the structural wood pieces will be completely missing. It is not uncommon to find roofs damaged because of many years of careless storage.

A good restoration job will include the removal and replacement of the fabric roof and the removal and replacement of the roof wooden structure.

If the car is a closed model, obtain a new fabric roof kit and install it yourself. These kits are available in two types, one is designed for coupes, the other is designed for sedans. They come complete with padding, roof material, hardware, and trim welt.

Wooden top bows are available from Aged-Auto Parts Co., Minneapolis, Minn., or you can obtain a complete top wood kit from Aged-Auto Parts Co. and install it yourself. This company has made available to the restorer complete top wood kits for coupes, sedans, and sport coupes. It is fortunate that at least this company is producing structural wood pieces because patterns for them are not readily available.

9.2 Removal of the old roof.

The metal moulding that goes around the edges of the fabric roof is very difficult to obtain, so use extreme care in removing it. Try to remove the molding without damaging it. However,

70

if it is missing or damaged to the point where it can not be reused, there is a wire-on type of molding available that can be used if necessary.

After the molding has been removed, remove the old fabric and all the nails.

If the outside body metal between the edges of the roof molding and the gutters is rough, sand or grind this area until it becomes smooth enough to refinish. The metal area that butts against the fabric should be smoothed down now rather than by a body man after the roof has been installed.

9.3 Installation of new roof.

Examine the condition of the wooden side rails and cross bows. If they have been replaced, leave them alone. If they are deteriorated, remove the metal fittings from each corner, remove all the roof hardware, and remove the wooden pieces. Figures A-1 through A-5 in the APPENDIX shows the location of each piece of wood in the roof structure of a variety of models.

Install the new top wood kit.

Install the fabric roof kit as follows:

1. Paint the inside curvature of the top of the metal body with two coats of rust-proof black enamel.

2. Stretch a piece of screen mesh over the roof bows from front to rear and staple it to the bows.

3. Lay a block of cotton wadding on top of the screen mesh, trim it to fit, and remove it.

4. Apply a coat of ordinary roof cement on all flat surface tack strips (side rails and bows) and lay on the cotton wadding.

5. Center the roof material and draw it tight front to rear using temporary tacks. See Figure 9-1.

6. After the centers have been secured, draw the material out to each corner, making sure that all wrinkles are removed.

7. Tack the four corners down. If the indications are that the top material will come out smooth and fairly tight, the temporary tacks should be driven in and more added as the material is pulled towards the outside edges.

8. Replace the metal molding along the edges of the fabric and the metal body curvature. Start at one end and work across, making sure there are no humps.

9. Paint the molding with a thick linseed oil type of paint. When the job is finished, clean the new roof surface with a mild soap and warm water.

71

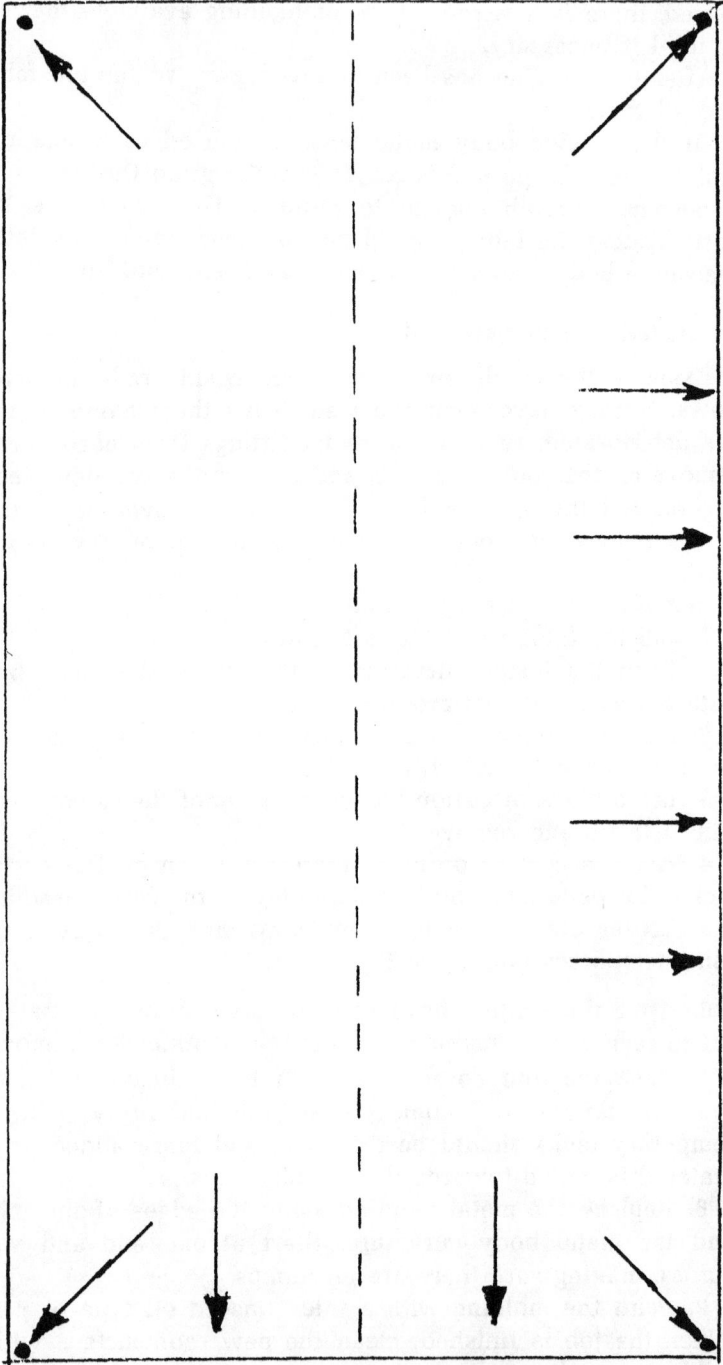

NOTE: Tack the corners temporarily, then work out from the center. The arrows show direction of pull.

Figure 9-1. Drawing the Roof Fabric

9.4 Original roof material.

All closed cars except Sport Coupe and Cabriolet: Black rubberized interlined.

Sport Coupe: Two-tone brown-gray, rubber, interlined, pyrexylin coated.

Cabriolet: Tan fabric interlined with rubber.

Phaeton: Black rubberized top material.

CHAPTER 10

INSTALLATION OF HYDRAULIC BRAKE SYSTEM

10.1 General.

The information contained in this chapter is presented to those who possess a Model A needing a complete new set of brakes or for those who have made the decision to incorporate the hydraulic method of braking. Even though being repetitious, it should be noted that absolute authenticity of the restoration will be lost when hydraulics are installed. However, there are many advantages to be gained by using hydraulic brakes and the restorer should seriously consider the pros and cons before proceeding or before abandoning the idea. The main and foremost reason for changing to hydraulics is SAFETY if the car is to be used for transportation purposes primarily. The reasons for and against the incorporation of hydraulic brakes are not detailed here for two reasons:

1. It is an individual decision to be made by each restorer, and 2. This book is being held to an instructive level, not one of controversy.

The first thing to do is to visit a junk yard and acquire a complete set of hydraulic brakes from a 1940 through 1948 Ford. Any set made during these nine years will do the job.

A complete set will include:

Two front backing plates (with hardware for shoes)
Two front wheel drums
Two front wheel cylinders
Two rear backing plates (with hardware and shoes)
Two rear wheel drums
Two rear wheel cylinders
One master cylinder

It is very important that several things be checked before leaving the junk yard, otherwise there will be complications later.

1. Make sure that each backing plate has a wheel cylinder on it and that they are the correct ones. The 1940 Ford front wheel cylinder diameters are marked 1¼ and 1-inch. The 1942 through 1948 front cylinders are 1⅜ and 1-inch, the rear cylinders are the same as 1939 through 1941.

2. Make sure that every backing plate has a pair of brass eccentrics at the bottom for positioning the brake shoes.

3. There should be a total of four strong springs to pull the brake shoes towards the wheel cylinders.

4. Each rear backing plate should contain an arm covering

each spring and two U-type keepers which are not part of the front assemblies.

5. Each rear backing plate should also contain an arm to which the end of the emergency brake cable is attached.

6. Any fittings, T's, or couplings which were used in the hydraulic steel lines should be removed and retained for possible future use.

7. If possible, retain the 16 sets of hardware used to secure the backing plates to the spindle plates as some of them may be used later.

8. If an equalizer is used on the emergency brake cable, retain it for possible future use.

Do not bother saving the steel lines or the emergency brake cable as new ones will be used later. Do not bother cutting out the brake pedal as the Model A brake pedal will be utilized.

Be sure that no parts are missing before leaving the junk yard as the smaller parts like rear backing plate brackets (over the springs) and the U clips are hard to locate and if forgotten, will cost money later. If the 1940 through 1948 car used small brackets near the front on each side frame member to secure one end of each flexible brake line, remove these also for future possible use. Check near the rear of the car and locate a third flexible line. If this line uses a bracket (or brackets), remove them for future use. Be sure that the rod coming out of the master cylinder (through the rubber boot) is there and not laying on the ground under the car after the master cylinder has been removed.

The original Model A wheels will fit the wheel drums of any 1940 through 1948 Ford. It is not recommended to use the parts from a 1939 Ford. All parts on the 1939 can be used with the exception of the drums. These will not fit the Model A wheels. This chapter is based on using the parts removed from a 1940 model.

A fair place to pay for the complete brake system is approximately twenty dollars. Again, be sure to get all the parts mentioned above because the cost will go up if parts have to be purchased later.

Load everything into boxes and take them home. The first thing to do is to acquire a small education.

Examine each backing plate and determine which one goes where. The front pair will appear to be identical, however, the direction of the larger end of each wheel cylinder will tell which plates goes on the front right wheel and which plate goes on the left front wheel. The larger end of the wheel cylinder goes on the front of each backing plate (nearest to the front bumper).

Examine the method used in mounting the brake shoes to the front backing plates. Note that the longer brake shoes face the front of the car. After becoming familiar with the location of each brake shoe, each front cylinder, and the brass eccentrics at the bottom of each assembly, identify and tag each eccentric and draw a sketch of each of their positions during the installation. Actually it is better to prick punch a number on each backing plate and on each eccentric so that the eccentrics can be returned to exactly the same position as original.

Disassemble the front units, clean the backing plates, and apply two coats of a rust-proof enamel.

Purchase a complete set of new brake shoes, four wheel-cylinder rebuilding kits, and a master cylinder rebuilding kit for the specific year of the new brake system (1940 through 1948).

Remove and retain all the parts from inside each wheel cylinder and from the master cylinder. Take the four wheel cylinders and the master cylinder to an automotive shop and have them honed. This is assuming that they are not deeply scored or broken and that the restorer does not possess a honing tool.

Replace the removed parts in the five cylinders with parts from the rebuilding kits and use the original pistons. Spread brake fluid around the honed areas with a finger and dip each part into brake fluid as you rebuild the cylinders. Store the cylinders until later when they will be installed.

Remove the original mechanical brake system from the car as follows. Pull everything off the four spindles. Remove all the service brake rods. Do not remove the foot brake pedal, the rod connected to it, or the service brake cross shaft. Do not remove the emergency brake handle, the rod connected to it, or the emergency brake cross shaft. These will be used in the new hydraulic system. Retain the four round grease baffles that are bolted to the spindle plates. Up near the front wheels the clevises which were pinned to the front brake rods may be sawed off as they will not be used and will rattle if left on. Do not throw away any parts until after the job has been completed and even then it does not seem right to dump any Model A parts. Someone else may have use for them.

At this point the four tapered spindles and the four spindle plates should be visible. Both the service and emergency brake linkages should be intact back to their respective cross shafts and the two front brake rod clevises will have been removed.

10.2 **Front brake work.**

Examine the front spindle plates. If there is a grease slinger on the outside center of the plate, remove both of them as they will prevent the wheel drums from seating correctly.

The next job is to install the front pair of backing plates on the front spindle plates. Obtain a set of front conversion adapter rings (four pieces). These are available from specialist suppliers and probably from local speed shops. You will see that the hole in the center of the front backing plates is too large and that the four mounting holes will not line up correctly with the four holes in the spindle plates. Patience here is necessary and the problems can be overcome.

Insert one of the larger adapter rings (3.625-inch inside diameter, 3.937-inch outside diameter, and $1/8$-inch thick) into the center hole of one of the backing plates and the backing plate will fit snugly against the spindle plate. Use a rat-tail file to elongate the four misaligned holes in the backing plate until they are aligned with the four holes in the spindle plate. Slide one of the original (Model A) circular metal grease baffles along the spindle and over the adapter ring. This will keep the ring in place. Bolt the assembly to the spindle plate and mount the correct wheel cylinder on the backing plate ($1^1/4$-inch piston towards the front if the cylinder came from a 1940 or 1941 Ford; $1^3/8$-inch piston towards the front if the cylinder came from a 1942 through 1948 Ford) with three bolts and lockwashers. (The hole for the steel tubing line will point towards the rear of the car).

Slide one of the smaller adapter rings (1.190-inch inside diameter, 1.560-inch outside diameter, and 11/32-inch thick) over the spindle with the beveled edge going on first. If it will not go all the way up to the spindle plate then it will be necessary to drive it on with a pipe and heavy hammer. It must go all the way on the spindle.

If new bolts and lockwashers are used to secure the baking plate to the spindle plate, drill holes in each bolt near the nuts and insert cotter pins. It may be possible to use some of the Model A bolts but not any from the 1940 assemblies.

Install the new brake shoes on the front backing plate with the longer one on the front. Install them exactly the same way as they were found on the car in the junk yard. Use brake pliers to install the springs. Turn the eccentric cam adjustments and watch the action of both shoes. A starting setting is to move each shoe **out** to its maximum horizontal forward position and vertically centered. Play with the adjustments for a while until familiar with their motion and action.

Remove any grease that may have accumulated on the brake shoes using steel wool.

Go to the other front wheel area and perform the same steps as detailed in the five preceding paragraphs.

Remove the front wheel bearings and the grease seals from the front wheels and new wheel drums.

If the bearings appear to be in good condition, use them on the new system. If not, replace them. Examine the wheel drum machined surfaces. If they are not scored from the old (1940-1948) brake shoes, use them on the new system. If they are scored (ridges dug into the machined surfaces), take them to a brake shop to be turned down. There is a safety factor involved here. Make sure there will be enough drum material left to maintain this safety factor.

If the original grease slingers have been removed and the backing plates are bolted up correctly, the wheel drums can be installed. If they will not clear the brake shoes, readjust the eccentric cams.

Flush out the centers of the wheel drums with kerosene, replace the inner and outer grease seals with new ones, pack the wheel bearings with bearing grease and install the bearings, install the wheel drums, securing them with the large slotted washer, nut, and a new cotter pin. Make sure that the axle key is in the end of each spindle. Install the dust covers. Bolt on the front wheels. Approximately 1/2-inch of the backing plate should be visible around the outer edge of the drums. Make sure that the openings in the wheel cylinders point toward the rear of the car.

The front end part of the conversion is now complete except for the brake fluid steel lines which will be installed after the rear wheel conversion has been completed.

10.3 Rear brake work.

Determine which backing plate goes on the rear right and which one goes on the rear left sides. Study them carefully so that after dismantling the units they can be correctly reassembled. The larger end of each wheel cylinder goes toward the front of the car and the longer brake shoe goes on the front. Also, the two openings to admit the emergency brake cable goes on the front of each backing plate.

Examine the method used in mounting the brake shoes to

the rear backing plates. Note that there is a bracket (or arm) covering the spring which is not part of the front wheel assemblies. After becoming familiar with the location of each brake shoe, each rear cylinder, and the brass eccentrics at the bottom of each assembly, disassemble the two rear units, and clean the two backing plates.

At this point it will be necessary to modify the rear backing plates because the ends of the rear spring will prevent the rear backing plates from snugging up tightly against the spindle plates.

Place a backing plate against the spindle plate and mark a rectangular box on the inside surface of the backing plate where the end of the spring makes direct contact with the backing plate.

Have this area cut out with a torch and then use a coarse file to get a smooth clearance fit. (See figure 10-1). After the plates have been fitted so they will install flat against the spindle plates, clean them up again and apply two coats of rust-proof black enamel.

NOTE: The four center mounting holes will not require elongating nor will any adapter rings be required on the rear backing plates.

Install the correct wheel cylinders on the backing plates (1^1/$_8$-inch piston towards the front) with three bolts and lockwashers. (The holes for the steel tubing lines will point towards the rear of the car). Bolt the backing plate assemblies to the spindle plates. Use longer bolts in each pair of front holes as they will also hold the radius rods in place. Drill holes in each bolt near the nuts and install cotter pins.

Install the new brake shoes on the rear backing plates with the longer ones on the front. Install the brackets (arms) across the springs and into the emergency brake arm. Don't forget the U-type clips. Use brake pliers to install the spring. Turn the eccentric cam adjustments as on the front and watch the action of both shoes. The starting setting is to move each shoe **out** to its maximum horizontal forward position and vertically centered. Play with the adjustments for a while until familiar with their action.

Obtain a new emergency brake cable for 1940. Feed it out of one backing plate entrance and into the other backing plate entrance tube. Snap the small balls on each end of the cable into the bottom of each emergency brake bracket on the backing plates. Pull the loop towards the front of the car and tie it

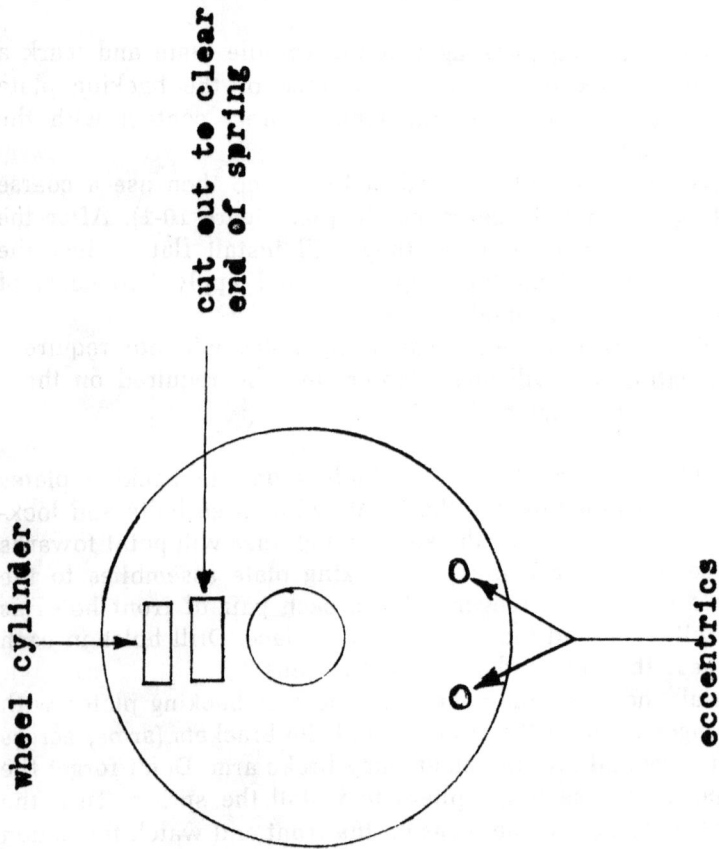

wheel cylinder

cut out to clear end of spring

eccentrics

Figure 10-1. Rear Backing Plate Modification

up with string until later when it will be connected to the emergency brake cross shaft.

Use steel wool to remove any grease that may have accumulated on the brake shoes.

Remove the grease seal and small outer wheel bearing from each wheel drum. If the bearings appear to be good, use them on the new system. If not, replace them. The wheel drums should be turned down, if necessary, by a brake shop, observing the safety factor. Replace the grease seals with new ones and pack the wheel bearings with bearing grease after flushing out the centers of the wheel drums with kerosene.

Before installing the wheel drums, it will be necessary to space the drum a few thousandths of an inch from the backing plates. Rear drum spacers are available or can be made from a piece of a tin can. Cut a piece of tin and bend it around the tapered spindles. It should start at one edge of the axle key, wrap around the spindle, and stop at the opposite edge of the axle key.

Install the brake drums, the slotted washers, and draw the nuts up as tight as possible. If the drums will not clear the brake shoes, readjust the eccentric cams. Install new cotter pins. Bolt on the rear wheels. Approximately $1/2$-inch of the backing plate should be visible around the outer edge of the drums. Make sure that the openings in the wheel cylinders point toward the rear of the car.

10.4 Master cylinder installation.

The master cylinder will be installed underneath the left side frame member on a rugged bracket and utilizing the holes in the frame occupied by the bolts securing the left end of the service brake cross shaft. This is an awkward place for adding brake fluid, but it simplifies the job and should be perfectly satisfactory after the installation has been completed. If the fluid lines are installed correctly, the addition of fluid will not present any problem. The reservoir level can be brought up by using a rubber hose to carry in the fluid or even a syringe could be used. A system that is really easy to fill will use a Saab reservoir mounted in the pan under the front seat. This is a plastic unit, complete with a heavy rubber hose, can be mounted easily, and will require an inlet connection to the master cylinder.

Remove the service brake cross shaft by removing the four clamp bolts and nuts. Remove the cotter and clevis pin from the rear end of the service brake rod. The left hand lever is to

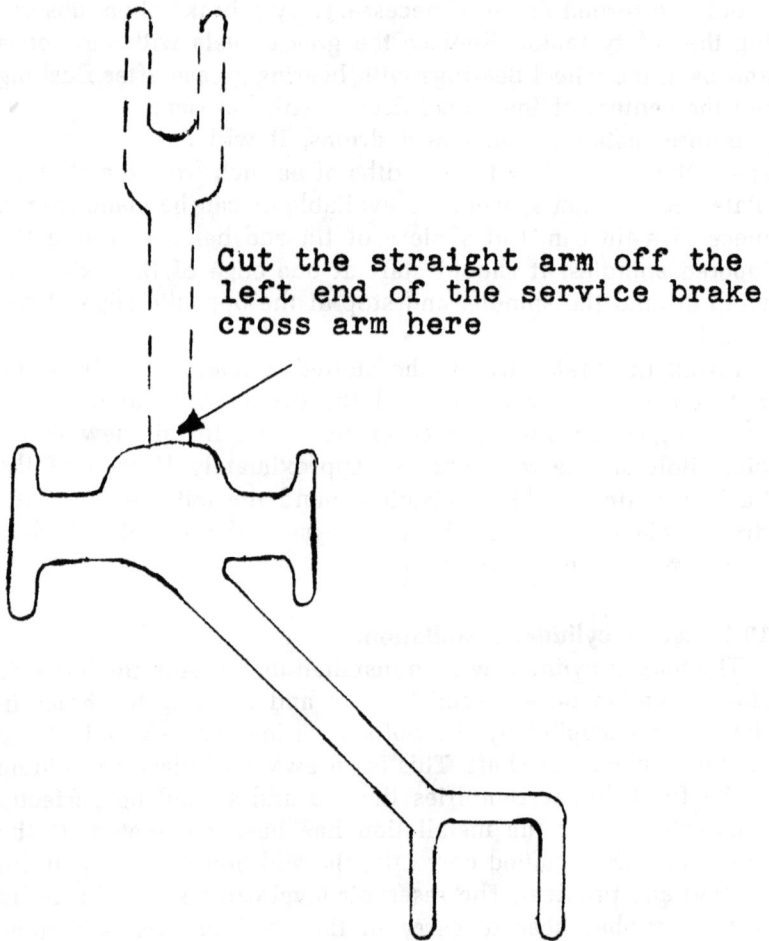

Cut the straight arm off the
left end of the service brake
cross arm here

Figure 10-2. Modification of the Service Brake Cross Arm

be modified next (see figure 10-2). Drill out the pin securing the left hand end lever to the left hand end of the cross shaft. Remove the lever, turn it 180 degrees and put it back on. Drive in a new pin and peen over the ends or insert a $^3/_8$-inch bolt, lockwasher, and nut. See Figure 10-2. Cut off the straight portion of the end lever and reinstall the cross shaft, leaving the left hand hardware loose. Reinstall the cotter and clevis pins and secure the brake rod to the cross shaft.

Obtain two pieces of $^1/_4$-inch thick steel, one $6^3/_4$ inches by $3^1/_4$ inches, the other $5^5/_6$ inches x $3^1/_4$ inches. Also obtain two triangular pieces $1^1/_2$ inches each side.

The pieces will be bolted together to form a bracket using No. 8 machine screws one inch long and lockwashers as shown in Figure 10-3 after cutting the holes dimensioned in the sketch.

Assemble the master cylinder to the bracket, making sure that all clearances are correct and then remove the cylinder. Arc weld or brake all the joints inside the bracket. Assemble the master cylinder on the bracket again and clamp the bracket in place underneath the left frame member. Scribe or mark the two bolt holes onto the bracket and remove the bracket. Drill these two holes with a $^3/_8$-inch drill and drill out the two holes in the frame with the same drill. Locate the area of the rivet head found underneath the frame onto the bracket and drill a $^3/_4$-inch hole for clearance. Clamp the bracket and the master cylinder underneath the frame, the brake shaft, its bracket and clamp are installed under the master cylinder bracket and the entire assembly is then secured to the frame with $^3/_8$-inch bolts, lockwashers, and nuts. Drill holes in the bolts and insert cotter pins. Adjust the length of the rod coming out of the master cylinder so that the master cylinder piston is completely retracted when the pedal is out. This adjustment may have to be trimmed up after the hydraulic pressure has been brought up.

10.5 Rebushing the pedals.
Examine the brake pedal for excessive side play. If it is sloppy, remove the pin from the left end of the clutch and brake shaft mounted on the transmission case, disconnect the clutch and brake pedal clevises and remove both pedals. Drive out the bushings from each pedal and insert new bushings. If the rod is worn, it will have to be replaced. If the pedals will not go back on the shaft, it will be necessary to have the

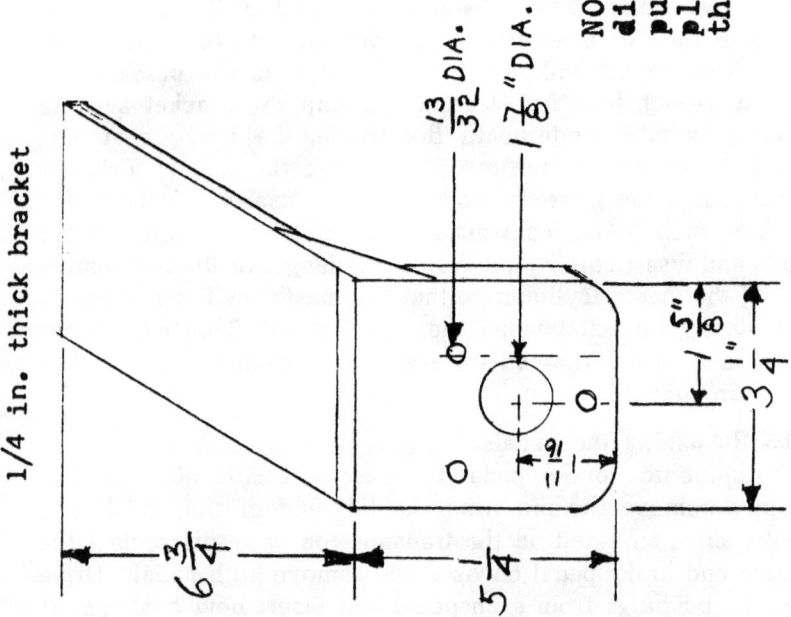

NOTE: Spacing of three 13/32-in. diameter holes is determined by putting the master cylinder in place and marking the 3 holes onto the bracket.

Figure 10-3. Master Cylinder Bracket

bushings reamed to the same diameter as the shaft ($^7/_8$-inch). Reinstall the pedals and their rods.

10.6 Installation of steel fluid lines.

Standard lengths of steel tubing can be used and no cutting or flanging is necessary.

Run a five foot length of $^1/_4$-inch steel line from each rear wheel cylinder along each rear radius towards the universal joint. Secure each line to each rear backing plate with a small clamp and attach three clamps to each radius rod.

It will be necessary to make up (if not previously salvaged) four metal brackets to secure the ends of three flexible lines. Two are used near the master cylinder, one at each end of an eight-inch flexible line. Use a rear flexible line made for a 1940 Mercury here. One of the brackets will bolt to a radius rod, the other to be bolted to the frame cross member. The remaining two brackets bolt on the side frame members using a hole existing for the securing of the engine splash pans. See Figure 10-4.

A complete layout of tubing lines, fittings, and cylinders is shown in Figure 10-5. Install the remaining $^1/_4$-inch steel lines, fittings, and flexible lines and shown and secure them whereever necessary with $^1/_4$-inch clamps.

Use the U-type clips (or keepers) at each front wheel angle bracket. The line between the two front wheel angle brackets should hug tightly against the front cross member and it should be kept clear of the front motor mounting stud.

10.7 Emergency brake work.

Connect the emergency brake cable to the emergency brake handle as shown in Figure 10-6. If the cable is too long, it can be shortened with a cable shortener available at automotive stores. The clevis and its shaft shown is modified from another Model A emergency brake rod.

10.8 Adjustment of the brakes.

An easy way to adjust each brake shoe cam is to place a wrench over each adjustment placing the wrench handle as close to horizontal as possible. Pushing the wrench down towards the ground will always move the brake shoes outward toward the wheel drum machined surfaces.

Adjust the cams in each wheel until the brake shoe hits the drum. Then back off the adjustment until the shoe clears the drum.

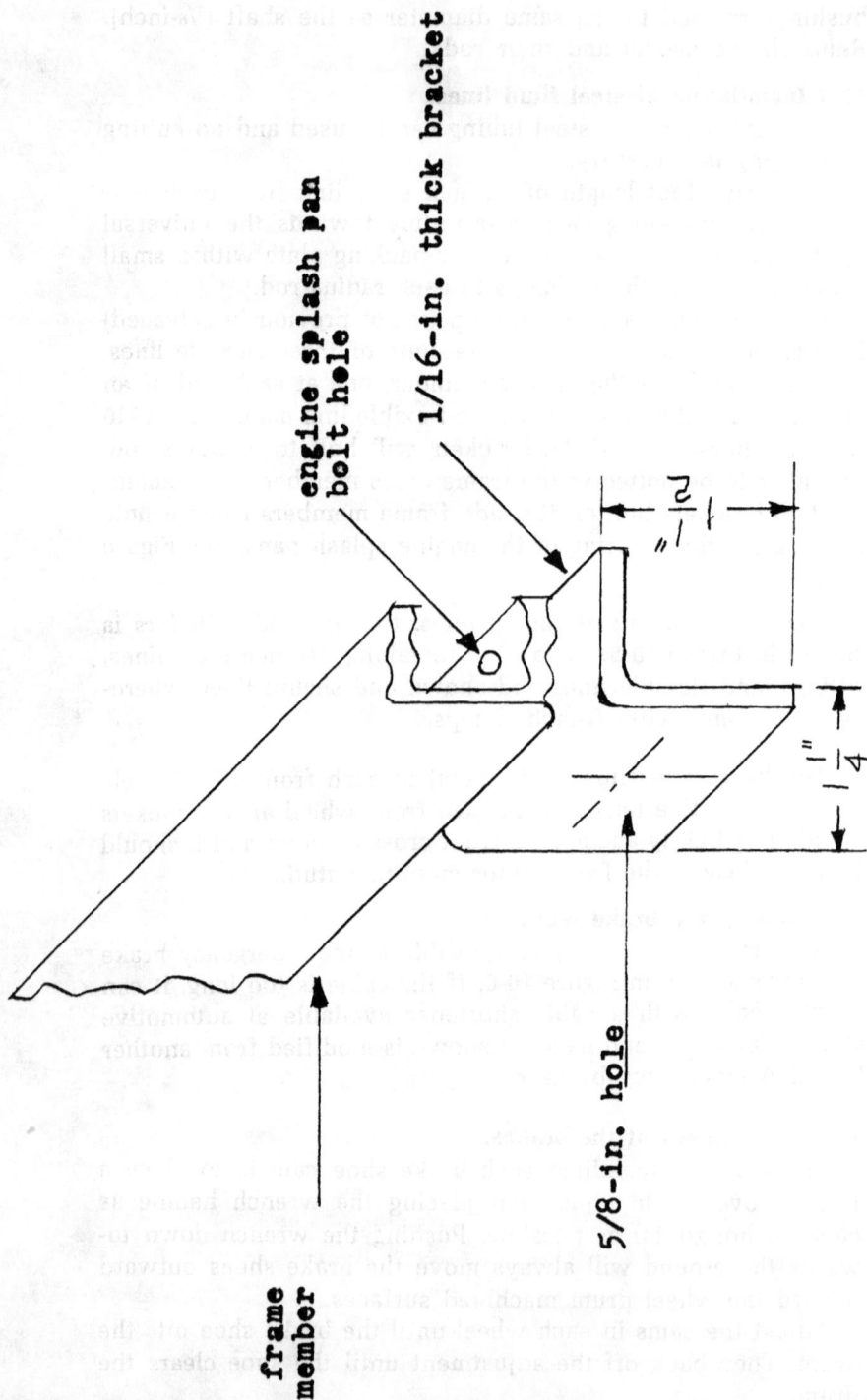

Figure 10-4. Location of Front Flexible Line Brackets

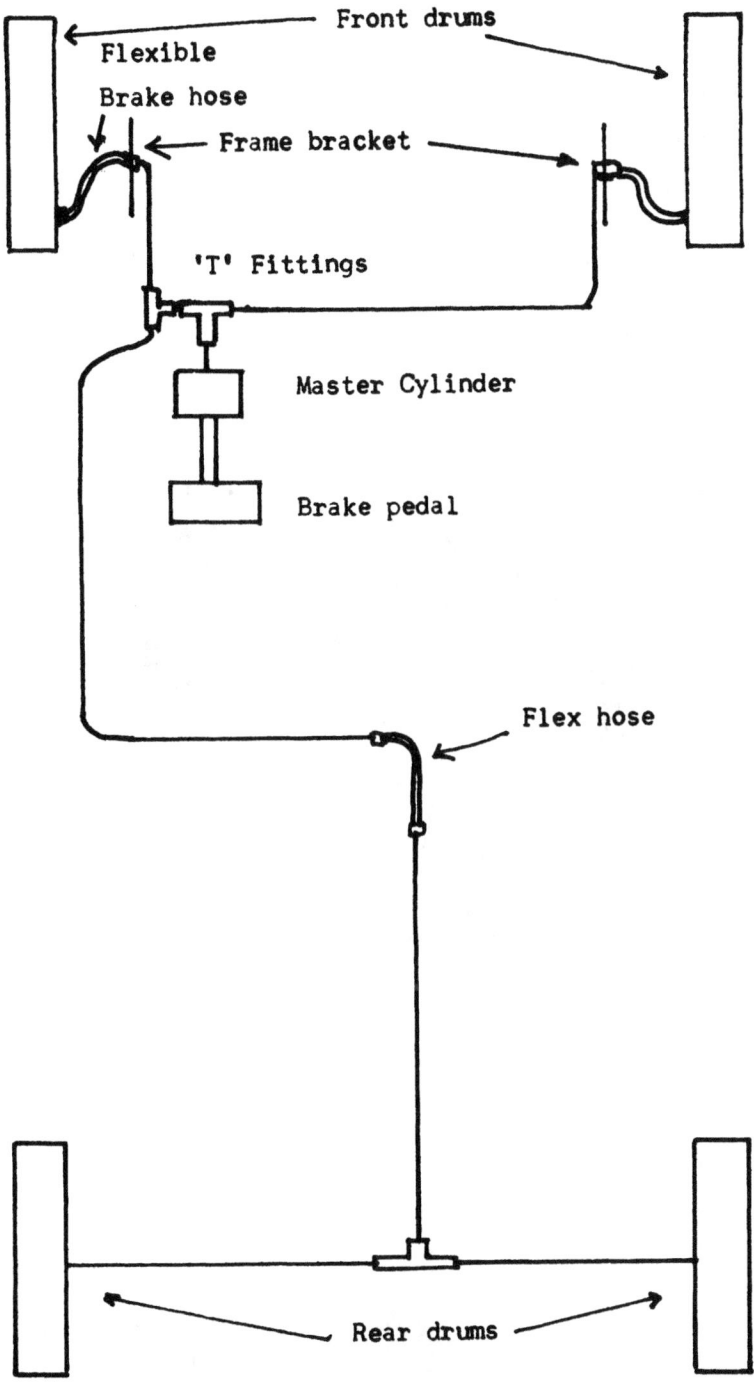

Front drums

Flexible
Brake hose

Frame bracket

'T' Fittings

Master Cylinder

Brake pedal

Flex hose

Rear drums

Figure 10-5. Hydraulic System Layout

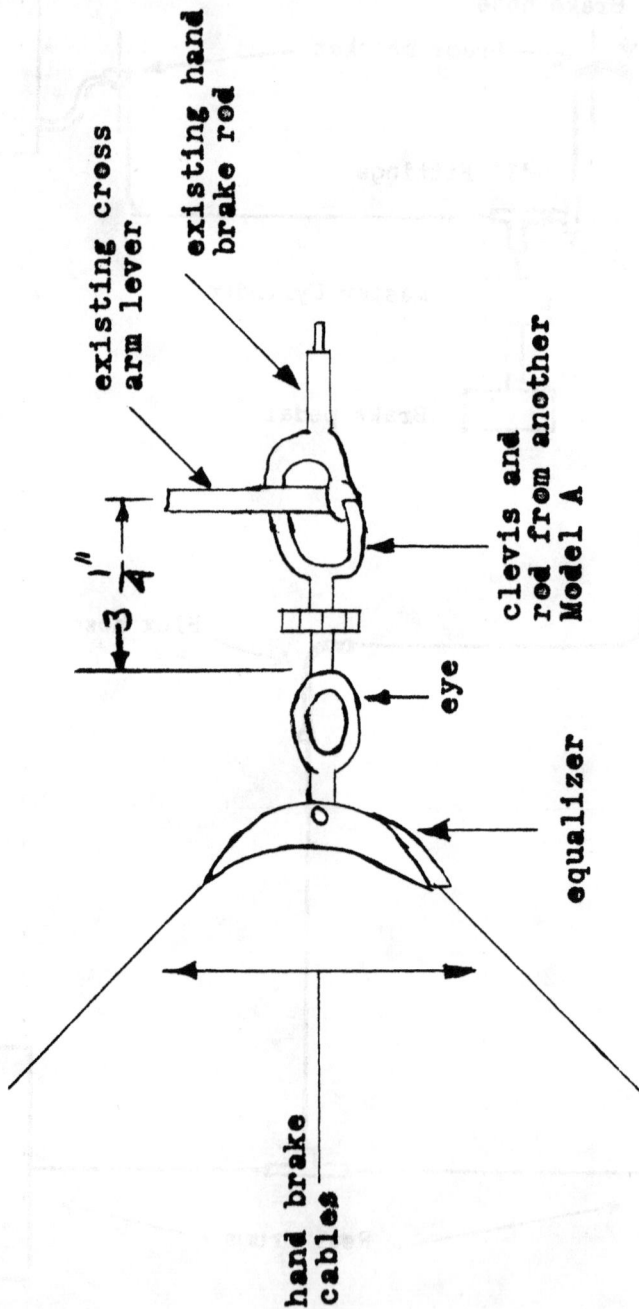

existing cross
arm lever

existing hand
brake rod

clevis and
rod from another
Model A

$3\frac{1}{4}"$

eye

equalizer

hand brake
cables

Figure 10-6. Installation of Emergency Brake Cable

10.9 Bleeding the hydraulic brake system.

Make sure that all joints and connections are tight. Remove the cover from the master cylinder reservoir and fill the system with a good high grade heavy duty brake fluid.

Loosen one of the wheel cylinder bleeder valves slightly. Have a helper pump the brake pedal until all air has been expelled. Catch the fluid in a can, or use bleeder hoses. The correct condition is arrived when no bubbles of air are expelled (fluid only). Push the brake pedal down slowly without stopping and tighten the bleeder valve as soon as the fluid starts to eject. Refill the master cylinder reservoir and check the bleed wheel area for leaks.

Perform the same steps on the remaining three wheels. It may be necessary to perform the whole bleeding process two or three times in order to obtain a good heard pedal action.

If it is impossible to get any fluid ejection from the wheel cylinders, loosen the line slightly at the master cylinder, pump the pedal and observe whether fluid ejects from the master cylinder. If it does, then there is a blockage somewhere in the line. If only one, two, or three wheel cylinders eject the fluid, then the ones that don't are defective. If the master cylinder does not eject fluid, remove it and check the holes in the bottom of the reservoir for rust blockage.

If it is impossible to obtain a good hard pedal before the pedal itself reaches the floorboard, it is possible that the fluid is leaking somewhere. Examine the wheel areas and all connections while the pedal is being worked. If no leaks are visible, remove the brake line from the master cylinder and insert a short bolt into the opening at the rear of the master cylinder. If a hard pedal is obtained now, the cylinder is good, and it will be necessary to bleed the wheels several more times.

When the bleeding process is finished, take the car to a Ford garage and have a mechanic make the final brake adjustments.

The two cams at the bottom of each set of brake shoes are to be adjusted so that each brake shoe will contact the brake drum evenly. This is difficult to do as it is impossible to see how well the shoes make contact with the drum. Ford garages have gages to measure the clearances around the machined surfaces of the wheel drums and it is for this reason that a well equipped garage make the final adjustments.

If the brake shoes move outward as they should, but with

no braking results, it is possible that the brake drums have been turned down too far and will have to be replaced with a better set of wheel drums.

CHAPTER 11

REWIRING THE ELECTRICAL SYSTEM

11.1 General.

1. Remove the battery if it is still in the car.

2. Remove the main wiring harness and any other wiring that may be found under the hood including the wiring that goes to the tail light, stop light switch, and to the instrument panel.

Rewiring will be covered in two segments:

1. Installation of the wiring necessary to start the engine.
2. Installation of wiring necessary for the operation of accessories.

11.2 Wiring necessary to start the engine.

1. Install a new terminal box on the firewall if the original one is broken, missing or in poor condition.

2. Obtain a new main wiring harness (try to find one that is correctly color coded), a new ignition switch with armored cable, the correct grommet for the terminal box, keys, a tail light crossover loom (to wire in a second tail light if your state law requires it), a terminal box-to-dash harness, a terminal box-to-generator harness, a cowl light harness, and the connection to be inserted between the coil and the distributor.

3. Remove the instrument panel and install a new ignition switch if it is necessary. Push the armored cable through the firewall, through the grommet in the terminal box, and secure it to the distributor (the internal connection is automatically made). This wire is a shielded one and has a connector which screws into the distributor. Connect the free end of the shielded wire to one contact on the ignition switch.

NOTE: All preliminary wiring refers to Figure 11-1.

4. Secure two wires of the terminal box-to-dash harness (yellow, and yellow with black tracer) to the ammeter. The opposite ends attach loosely to separate studs in the terminal box. Secure one end of the black wire to one terminal on the coil and the other end loosely to the stud in the terminal box which now has the yellow wire with black tracer attached to it. Attach one end of the red wire to the remaining terminal on the ignition switch and the other end to the remaining terminal on the coil.

5. Replace the cutout relay on the generator and the starter switch, if necessary.

6. Secure one end of the yellow wire in the terminal box-

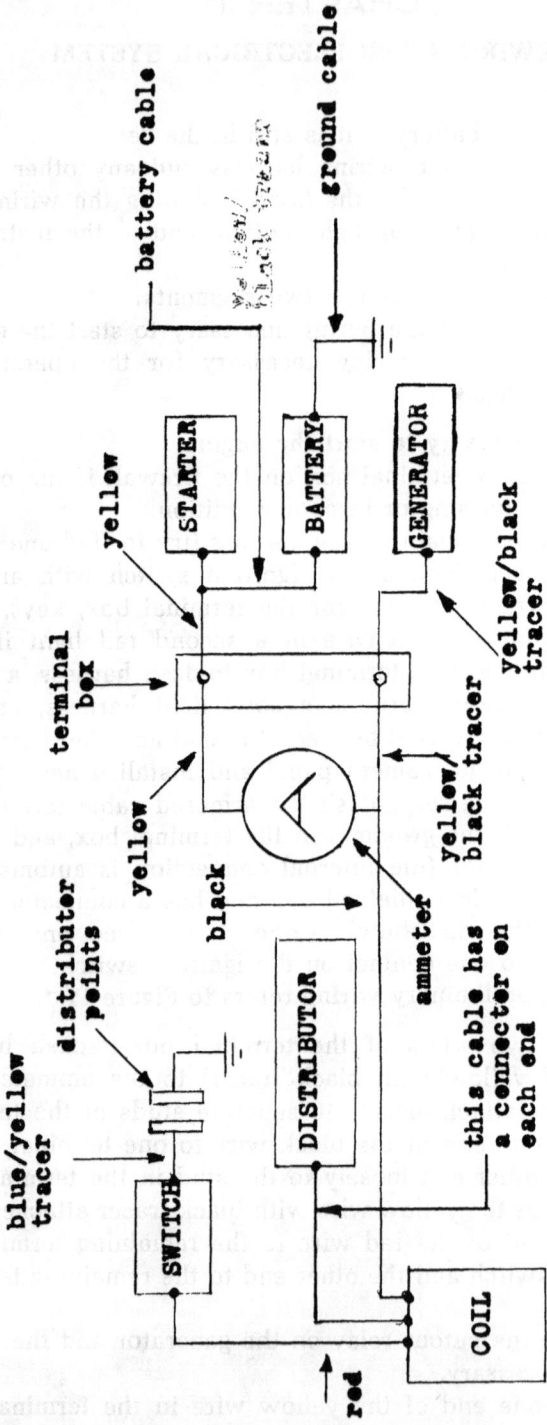

Figure 11-1. Preliminary Wiring

battery cable

ground cable

STARTER

BATTERY

GENERATOR

yellow

yellow/black tracer

terminal box

yellow

black

yellow/black tracer

ammeter

distributor points

blue/yellow tracer

SWITCH

DISTRIBUTOR

this cable has a connector on each end

COIL

red

to-generator harness loosely to the stud in the terminal box which has a yellow wire attached to it. Secure the other end of this wire to the large terminal on the starter motor. Attach the battery cable to the starter a this time. Check the battery ground cable at this time also and replace it if necessary.

7. Secure the end of the yellow wire with black tracer to the stud in the terminal box which has the yellow wire with black tracer attached to it. Secure the other end to the generator cutout relay.

8. Tighten both sets of connections inside the terminal box.

9. Tighten down the clamp on the firewall and the clamp on the left side of the engine block to secure the terminal box-to-generator harness.

Enough of the wiring system has now been installed to check out the operation of the engine, if desired.

11.3 Completion of electrical wiring.

1. Insert the free ends of the main harness into the metal harness housing, pull the harness through, and attach the housing to the metal light switch cover.

Attach the light switch/harness assembly to the bottom of the steering column securing it with the light switch spider bail.

3. Lay the main harness in position so that the green wires run towards the rear end of the car (underneath the front floorboards).

4. Attach the green wires having eyelets to the stop light switch.

5. Connect the black and green wires from the contact plate to the connections on the tail light.

6. Use small clips to secure these wires to the left hand frame member.

7. Pull the short black wire with green tracer and the short black wire with red tracer through the left headlight wiring hole in the radiator shell, slide a rubber grommet down to the shell, slide the headlight conduit over these two wires and attach them to the bright and dim connections on the left headlight respectively.

8. Extend the longer black wire with green tracer and the longer black wire with red tracer along the bottom of the radiator and pull them through the right headlight wiring hole in the radiator shell, slide a rubber grommet down to the shell, slide the headlight conduit over these two wires and

attach them to the bright and dim connections on the right hand headlight respectively.

9. Secure the right headlight wires with small clips.

10. Pull the blue wire with yellow tracer through the remaining wiring hole in the radiator shell, pull one end of the yellow wire through the hole, slide a rubber grommet down to the shell, slide the horn conduit over these two wires, and attach them to the horn connections.

11. Attach the black and white wire with a connector on it to the generator cutout relay. Be sure to use the small clips to secure the wires to the frame wherever necessary.

12. The black wire with blue tracer goes from the terminal box to the windshield wiper as shown in Figure 11-2.

13. Another black wire should be run between the ammeter and the instrument panel light as shown in Figure 11-2.

NOTE: Two main wiring harnesses are available. One is for use on a car with front parking lights and one is to be used on cars having cowl lamps. A separate cowl light harness is necessary for cars with cowl lights and is to be used with a main harness designed for cars having cowl lights.

14. If the car uses cowl lights, install the cowl light harness and connect it to the two black wires with yellow tracers which have been broken out of the main harness.

15. If state law requires two tail lights, use a tail light cross over harness and connect it in parallel with the existing tail light wiring.

The use of sealed beam headlights will destroy the authenticity of the restoration. However, for extended night driving they are superior and safer. Sealed beam adapter kits for the Model A are available. The current carrying capability of the harness will be taxed by the use of sealed beams as they require much more current than the original bulbs.

Figure 11-2 is a diagram showing the connection of all units, Figure 11-3 is a diagram for cars not equipped with cowl lights, Figure 11-4 is for cars equipped with cowl lights and Figure 11-5 is a pictorial of the electrical system. Figure 11-6 is the Fordor wiring diagram, and Figure 11-7 is a wiring diagram of the Town Sedan, standard Fordor sedan, and DeLuxe coupe.

11.4 **How to determine if the horn is authentic.**

It is possible that unknowingly a horn may be obtained which is not original for the Model A. There are some that sound like Model A horns but in actuality are not.

Originally, the horns were supplied by Sparton, E.A., and

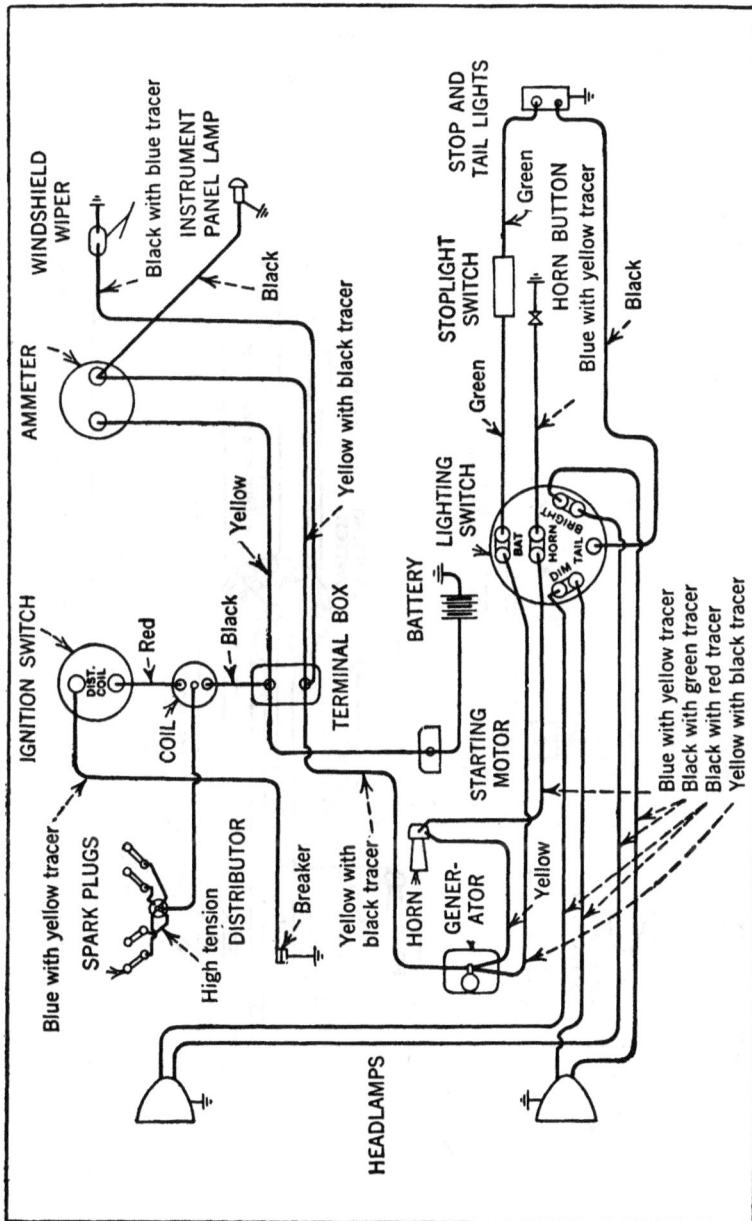

Figure 11-2. Wiring Diagram of Electrical System Showing All Units

WINDSHIELD WIPER

Black with blue tracer

INSTRUMENT PANEL LAMP

Black

AMMETER

Yellow

Yellow with black tracer

IGNITION SWITCH

Red

Black

COIL

DIST. COIL

TERMINAL BOX

BATTERY

Blue with yellow tracer

SPARK PLUGS

High tension

DISTRIBUTOR

Breaker

Yellow with black tracer

HORN

GENER-ATOR

STARTING MOTOR

LIGHTING SWITCH

Green

BAT

HORN

BRIGHT

DIM

TAIL

Yellow

Blue with yellow tracer
Black with green tracer
Black with red tracer
Yellow with black tracer

HEADLAMPS

STOP AND TAIL LIGHTS

Green

STOPLIGHT SWITCH

HORN BUTTON

Blue with yellow tracer

Black

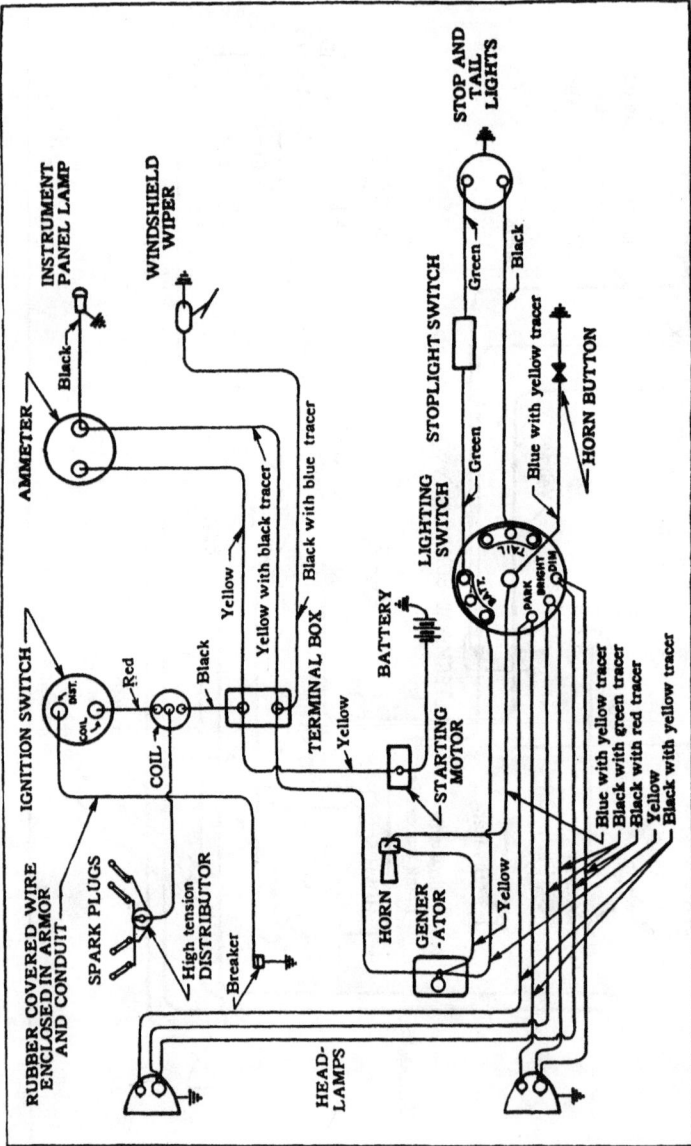

Figure 11-3. Wiring Diagram for Cars Not Equipped with Cowl Lights

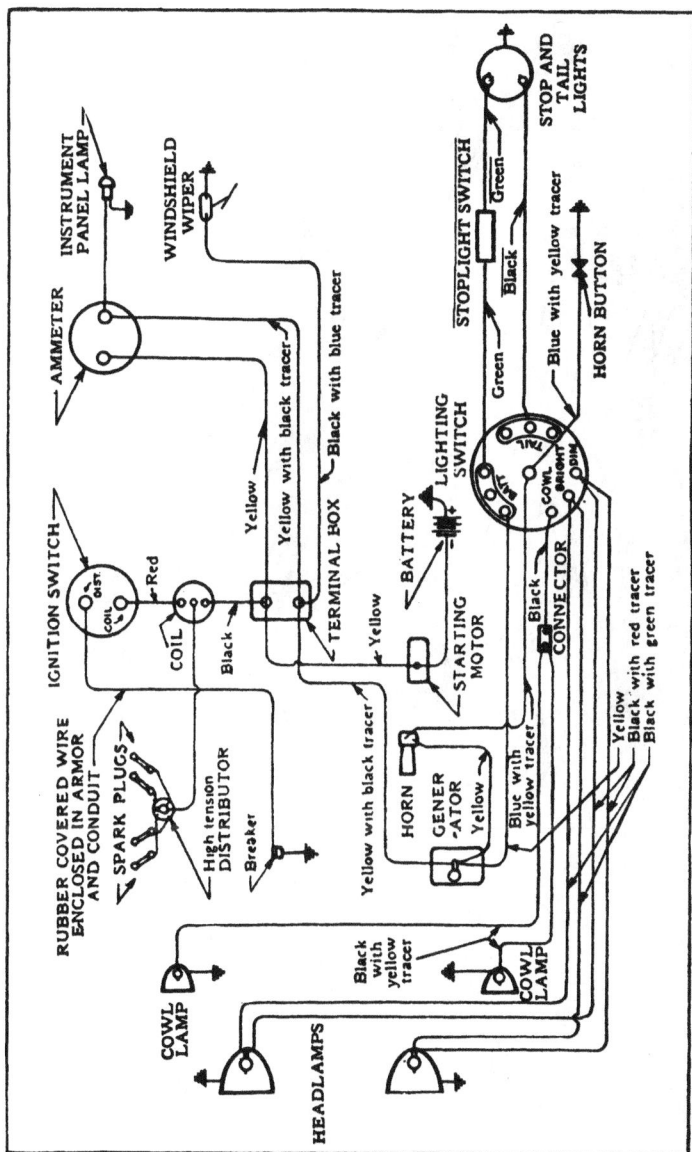

Figure 11-4. Wiring Diagram for Cars Equipped with Cowl Lights

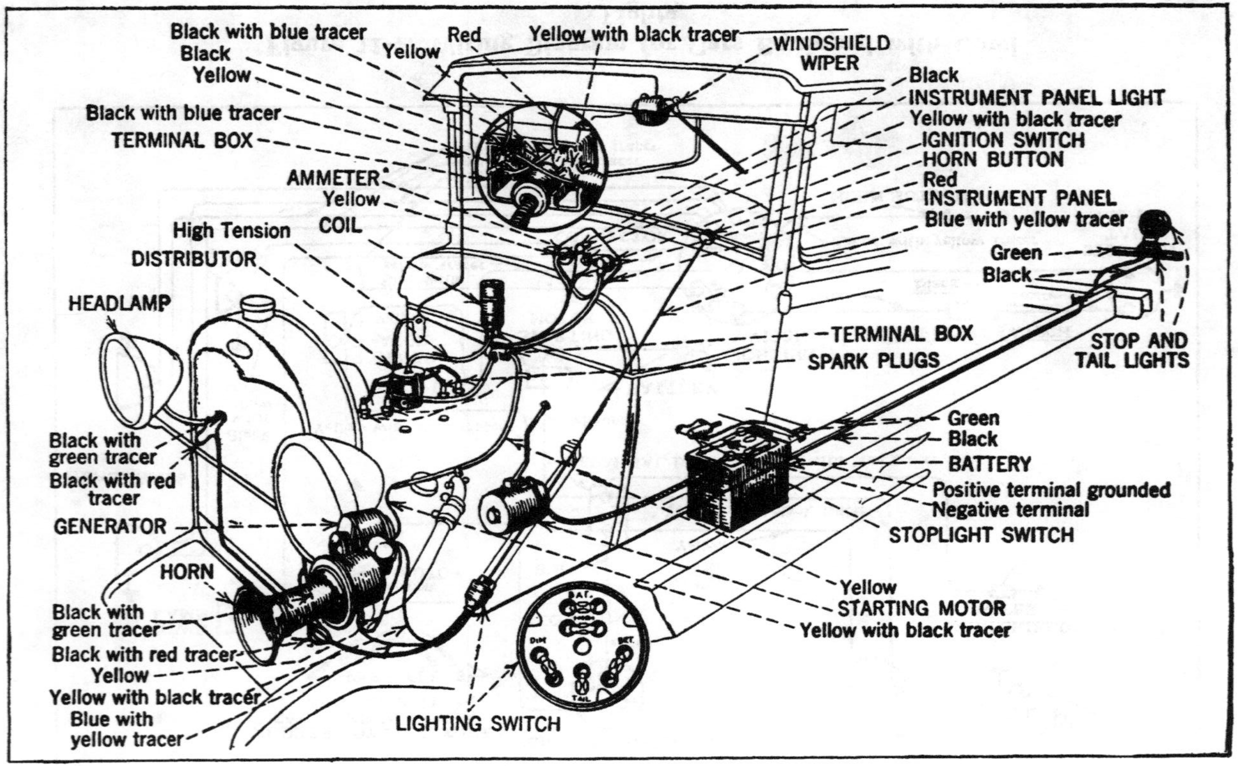

Black with blue tracer
Black
Yellow
Black with blue tracer
TERMINAL BOX
AMMETER
Yellow
COIL
High Tension
DISTRIBUTOR
HEADLAMP
Black with green tracer
Black with red tracer
GENERATOR
HORN
Black with green tracer
Black with red tracer
Yellow
Yellow with black tracer
Blue with yellow tracer

WINDSHIELD WIPER
Yellow with black tracer
Red
Yellow
Black
INSTRUMENT PANEL LIGHT
Yellow with black tracer
IGNITION SWITCH
HORN BUTTON
Red
INSTRUMENT PANEL
Blue with yellow tracer
TERMINAL BOX
SPARK PLUGS
Green
Black
STOP AND TAIL LIGHTS
Positive terminal grounded
Negative terminal
BATTERY
Green
Black
STOPLIGHT SWITCH
Yellow
STARTING MOTOR
Yellow with black tracer
LIGHTING SWITCH

Figure 11-5. Pictorial Diagram of Electrical System and Part Locations

98

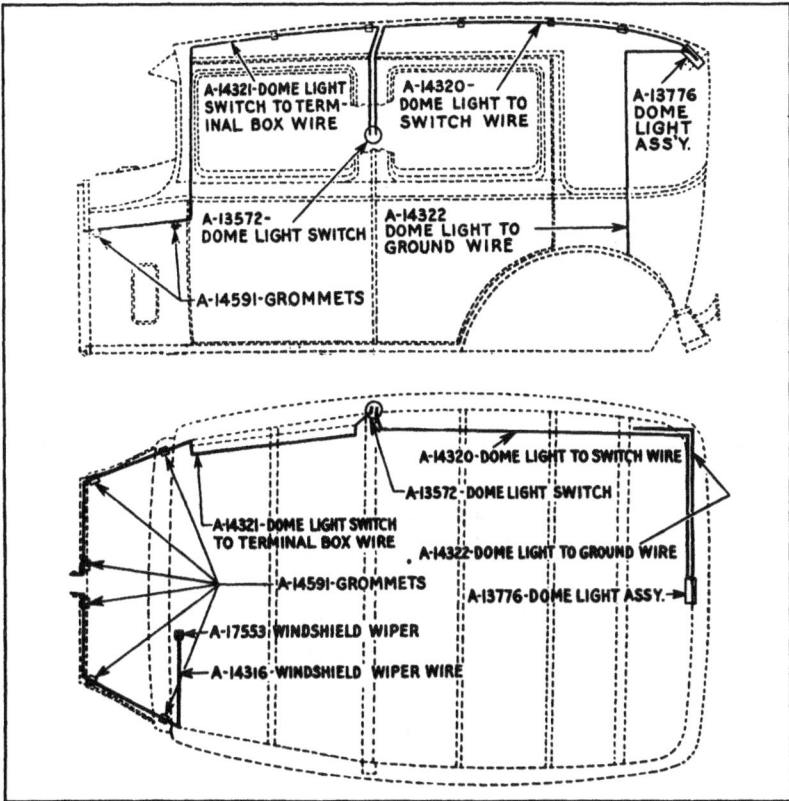

Figure 11-6. Fordor Wiring Diagram

Ames. Determine if the horn is original and determine which one it is by removing the cover and measuring the distance from the adjusting screw to the hole for the horn cover screw. The dimensions should be as follows:

E.A. — 17/32-inch

Sparton — 1¾₆-inches

Ames — 1-inch

Attach a new patent plate to Sparton horns if the original one is missing or deteriorated.

11.5 **Repairing the horn rod.**

If the wire inside the horn button rod is broken, it will have to be repaired. Unsolder the small contact from the center wire at the bottom of the horn rod, being careful not to lose the small contact and bushing. Bend the tabs that secure the rod to the button **very carefully.** Remove the wire from the button. Install a new length of stranded wire inside the rod, soldering

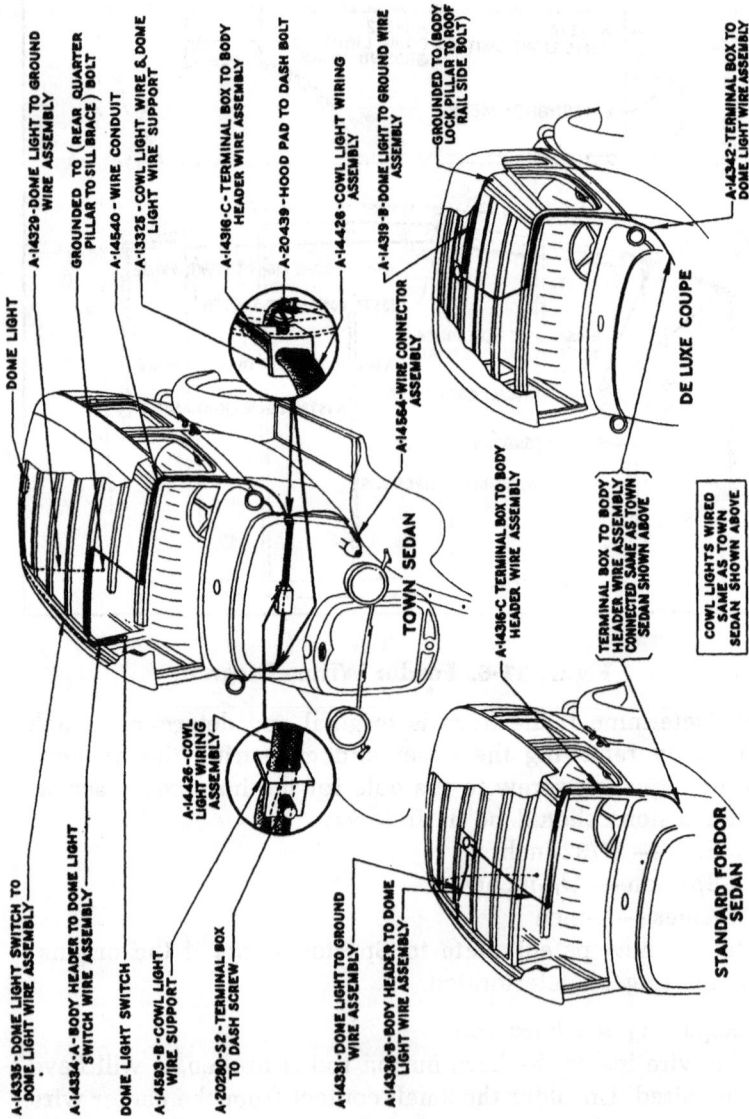

A-14335 - DOME LIGHT SWITCH TO DOME LIGHT WIRE ASSEMBLY

A-14334-A - BODY HEADER TO DOME LIGHT SWITCH WIRE ASSEMBLY

DOME LIGHT SWITCH

A-14583-B - COWL LIGHT WIRE SUPPORT

A-20250-S2 - TERMINAL BOX TO DASH SCREW

A-14331 - DOME LIGHT TO GROUND WIRE ASSEMBLY

A-14338-B - BODY HEADER TO DOME LIGHT WIRE ASSEMBLY

A-14426 - COWL LIGHT WIRING ASSEMBLY

DOME LIGHT

A-14329 - DOME LIGHT TO GROUND WIRE ASSEMBLY

GROUNDED TO (REAR QUARTER PILLAR TO SILL BRACE) BOLT

A-14540 - WIRE CONDUIT

A-13325 - COWL LIGHT WIRE & DOME LIGHT WIRE SUPPORT

A-14316-C - TERMINAL BOX TO BODY HEADER WIRE ASSEMBLY

A-20439 - HOOD PAD TO DASH BOLT

A-14426 - COWL LIGHT WIRING ASSEMBLY

A-14319-B - DOME LIGHT TO GROUND WIRE ASSEMBLY

GROUNDED TO (BODY LOCK PILLAR TO ROOF RAIL SIDE BOLT)

A-14342 - TERMINAL BOX TO DOME LIGHT WIRE ASSEMBLY

A-14564 - WIRE CONNECTOR ASSEMBLY

TOWN SEDAN

DE LUXE COUPE

A-14316-C TERMINAL BOX TO BODY HEADER WIRE ASSEMBLY

TERMINAL BOX TO BODY HEADER WIRE ASSEMBLY CONNECTED SAME AS TOWN SEDAN SHOWN ABOVE

COWL LIGHTS WIRED SAME AS TOWN SEDAN SHOWN ABOVE

STANDARD FORDOR SEDAN

Figure 11-7. Wiring Diagram of Town Sedan, Standard Ford Sedan, and DeLuxe Coupe

it to the button. Reassemble the button to the rod carefully so as not to break the tabs. Insert the small bushing ino the bottom end of the rod with the wire coming through its center hole. Slip the small contact over the end of the wire, up against the bushing, and solder the wire to the contact.

CHAPTER 12

INTERIOR OF THE CAR

12.1 General.

To restore the interior of the car correctly it is necessary to remove everything first that can be removed. This will include the battery, the upholstery if poor, the rubber boot and knob from the gearshift handle, the accelerator pedal, floor mat, floorboards, seat cushions, and front seat metal frame. Clean and paint (Ford engine green) the transmission and large housing if not already done. Remove the eight nuts that secure the front seat framework to the two runners underneath the seat frame. Remove the framework, clean it, and apply two coats of rust-proof black enamel. Remove the two runners, clean the old grease from the tracks, lubricate them with new grease, and reinstall the runners.

Remove the two identification plates found under the front doors. Obtain a new set if necessary, and install them after performing the next step.

Sand and clean the body sill areas, the metal that holds the floorboards in place, the battery box, and the inside of the firewall. Apply two coats of rust-proof black enamel.

Install a battery cable support kit on the clutch housing.

12.2 Work in the battery box.

If the bottom plate of the battery box is bad, replace it. These are available or can be made up. Cut a piece of 1/4-inch thick aluminum 7 by 9 inches. Clamp the plate in place and working from the underside, mark the two mounting holes. Remove the plate, drill the two holes, and remount it using countersunk bolts, nuts, and lockwashers. The commercial plate will have a water groove in it. Apply two coats of rust-proof black enamel to the plate. Clean up the battery post clamps and paint them, or replace if necessary.

12.3 Repair or replacement of the curtain pan.

Examine the condition of the metal pan directly under the front seat. If there are holes in the bottom of it, drill out the rivets and remove the pan. Obtain a replacement curtain pan (Palmer's Reproductions), or have a new one made in a sheet metal shop using the removed one as a pattern. The replacement pan will have the original grooves in it whereas the one from a local sheet metal shop will not. Secure the new pan in place with machine screws, nuts, and lockwashers. Apply two coats of rust-proof black enamel to the pan.

12.4 Package tray (coupe).

If there is any upholstery left on the package tray behind the seat, remove it and save the cardboard for reupholstering use later on.

While working in this area, now is the time to replace the rear window glass channels if they have deteriorated. Remove the package tray (coupe), the bezel around the rear window, and any wood which may prevent the removal of the glass. Remove the glass, the channels, and the rubber along the top edge of the glass.

Install new rear window channels, the glass, any wood which may have been removed, the bezel, and the package tray (coupe). Apply two coats of rust-proof black enamel to the package tray.

12.5 Floorboards.

Reinstall the original floorboards only if they are in good condition. During the four years of Model A production, two types of floorboards were used so be sure to obtain the correct set for the particular year car being restored. The floorboards used in the 1930-1931 models are different than those used in the 1928-1929 models. Plywood floorboards are available from several suppliers or they can be made if the facilities are available. Use ³/₄-inch thick marine plywood and see Figure 12-1 for the 1930-1931 dimensions.

Work them over a bit to obtain a good fit. Drafts and squeaking can be prevent to some extent by installing a felt edge around the two floorboard pieces. Apply two coats of porch and deck enamel to the floorboards. Kits of floorboard mounting hardware are available from several sources. Do not install the floorboards yet.

Have the gearshift handle and emergency brake handle chrome plated or nickel plated, as applicable, if necessary, before installing the floorboards.

12.6 Instrument panel.

Remove the instrument panel and cowl strip. Clean, sand. and paint (two coats of rust-proof black enamel) the cowl strip and the cowl areas. If the ends of the cowl strip are broken, have them brazed back into place. Install the cowl strip using new trim screws. Have the instrument panel replated if neces-

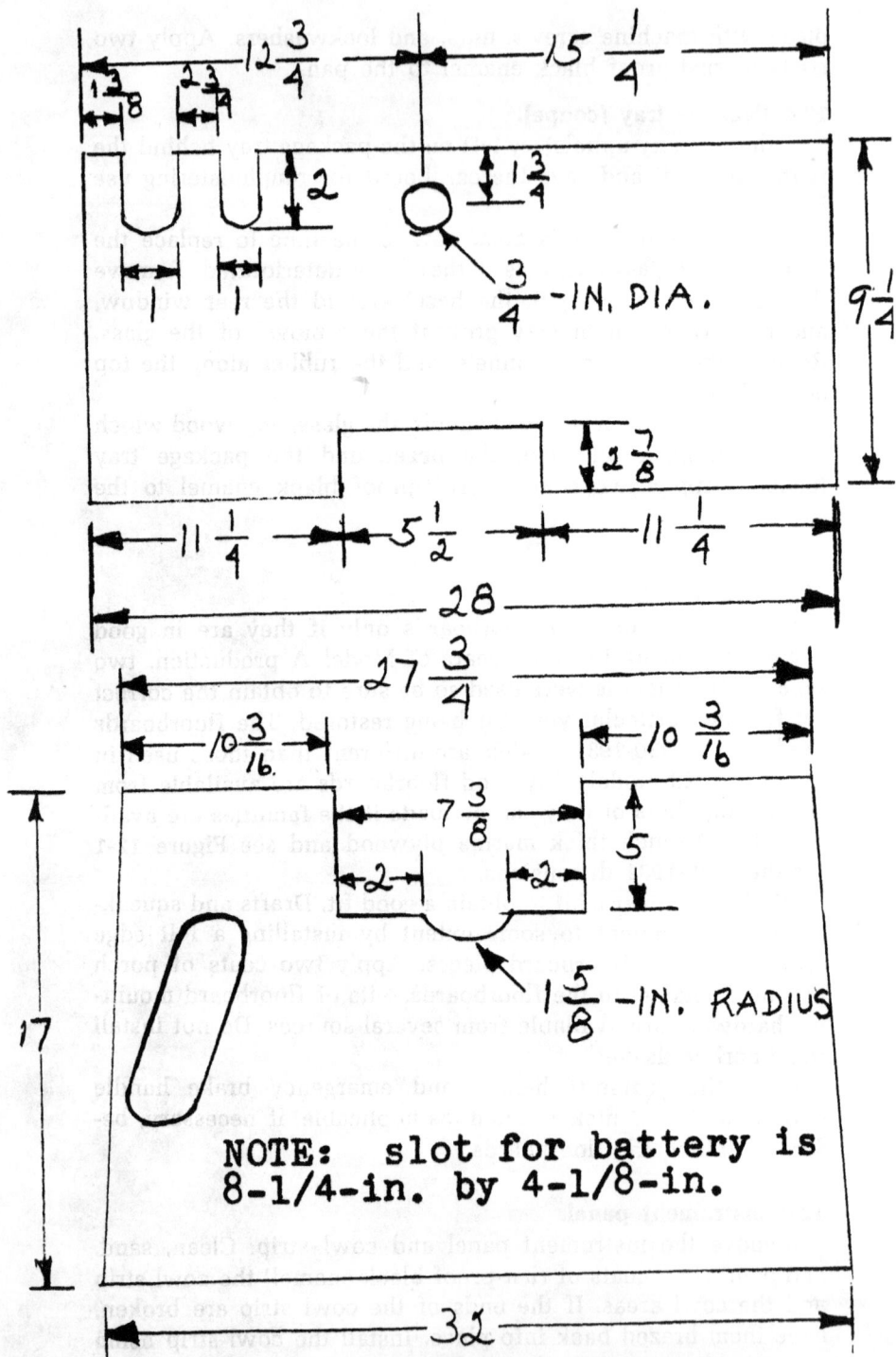

Figure 12-1. 1930-1931 Floorboard Dimensions

sary, and secure it with new trim screws. Replace the speedometer numerals with a new decal set if necessary.

12.7 Installation of the remaining units.

1. Reinstall the front seat framework and secure it to the metal slides with the eight nuts previously removed, or replace them if necessary.

2. Install the seat cushions.

3. Install a new grommet on the starter rod, two new grommets on the choke rod, and a new grommet on the speedometer cable, all at the firewall.

4. Install the floorboards, the floorboard battery cover and a steering post felt pad on top of a new rubber front floor mat.

5. Screw on the round accelerator pedal.

6. Install new rubber covers on the clutch and brake pedals, the starter switch and the accelerator pedal.

7. Install a new steering column rubber at the bottom of the cowl.

8. Install a foot rest adjacent to the accelerator pedal if it is missing and put a new rubber cover on it.

9. A complete job will include the removal and reinstallation of the gas tank using new gas tank welt available as a kit.

10. Refinish the mirror if necessary. The correct glass dimensions are $2^1/2$ by 6 inches.

11. Replace inner and outer door handles, if necessary.

12. Replace window winder handles, if necessary. If the original door and window handles are retained, have them nickel plated (1928-1929) or chrome plated (1930-1931), if necessary.

12.8 Glass repairs or replacement.

1. Remove the front windshield and install a new windshield rubber, closed car or open types. There are two different types used in the open cars, one for 1928-1929 and one for 1930-1931. If the car being restored is a closed model (any year), install a new header seal.

2. Polish the windshield swing brackets.

3. Remove the door and quarter panel windows, if necessary, and replace the channels.

4. Install new windlace door welting around the doors and their openings as was done originally.

12.9 Doors.

1. If the doors do not close properly, examine for sprung

hinges and replace them if necessary. Each door has a movable bracket in its jam to permit adjustment of the door closing.

2. Install a new door lock spring kit, if necessary.

3. Install a set of door check straps and brackets to prevent future springing of the hinges.

4. Install a set of rubber door bumpers. These are available for each different model and are part of a special car set.

5. Install a door arm rubber check on each door.

6. Apply two coats of rust-proof black enamel to all door edges and the openings inside each door.

CHAPTER 13

REUPHOLSTERING

13.1 General.

It is best to approach the reupholstering phase on a professional level which means purchasing the upholstery materials and having the work done by a professional automotive reupholsterer. A good man, when furnished with the correct materials can make the interior of a Model A a thing of beauty.

13.2 Headlinings.

Mohair material is used in most models and should be used to cover the roof section, the panels above each door, the headboard above the windshield, above and below the rear glass, and the quarter panels. The amount of material necessary to cover these areas should be ascertained and obtained. Several grades of materials are available and one of the specialist firms can supply materials which is very close to the original types including the mohair material.

Along with the covering, a trim kit should be obtained consisting of mouldings, adhesive, and tacks.

13.3 Door and side panels.

The cardboard pieces originally used as a backing for the door and side panels can be used as patterns for new ones if they came with the car. If not, the upholsterer can make new ones to fit the doors and the kick panel areas. These should be covered with new material and can be made from fibreboard.

13.4 Seat reupholstering.

If the seat upholstering is to be restored, most models used mohair covering and work should be done by a professional. Do not use many material except that which is very close to the original. Rumble seats were done in leather as were roadster front seats. The rumble seat kick panels were also covered with leather.

Stitt's has do-it-yourself literature available to aid in estimating the amount of material needed and instructions on how to cut a floor mat or carpet are also available.

13.5 Original types of upholstery.

If the reupholstering is to be done locally, be sure that the correct material is obtained.

PHAETON: Two-tone cross cobra grain artificial leather.

ROADSTER: Two-tone cross cobra grain artificial leather, rumble seat, same.

13.6 Interior trim.

Mohagany garnish moldings are found in the DELUXE COUPE, DELUXE SEDAN, TOWN SEDAN. Silk curtains on the rear quarter windows and a flexible robe rail are found on the TOWN SEDAN. The robe rail is also a feature of the DELUXE SEDAN.

COUPE: Brown check wool cloth. Pockets in doors.

DELUXE COUPE: Mohair, green or taupe. Bedford cord, deep tan.

SPORT COUPE: Above belt line, tan cloth. Below belt line, brown check. Rumble seat, two-tone cross cobra grain artificial leather.

TUDOR SEDAN: Brown check cloth.

CABRIOLET: Tan Bedford cord upholstery and trim below belt line. Rumble seat, two-tone cross cobra grain artificial leather.

DELUXE SEDAN: Tan Bedford cord, taupe or green mohair.

TOWN SEDAN: Brown or green mohair.

STATION WAGON: Two-tone cross cobra grain artificial leather.

CHAPTER 14

REFINISHING THE EXTERIOR SURFACE

14.1 General.

The outside surfaces of the car should not be painted until after all restoration work has been completed and after it has been ascertained that the car is functioning perfectly. Any malfunctions should be corrected before painting the car.

Only a professional should paint the car, generally speaking. However, there are certain things which can be done before taking the car to a body shop.

14.2 Preparation of the car for painting.

The original color combinations used on the Model A's has appeared previously in print but are given here again so that the authenticity of the colors will be retained.

In addition to the color information contained in this chapter, there is a Paint Manual on the market which includes chips of the original paint colors.

Determine which authentic color or colors are going to be used and obtain the paint (lacquer). After the surfaces have been prepared, take the car and the paint to a body shop and have it painted by a professional. Regardless of how good a job has been done in restoring the car, it can be ruined by a poor paint job. Be sure to obtain the paint rather than depending on the body shop for it, thus retaining the authenticity.

Remove the hood from the engine. Clean the inside surface with sandpaper or use a paint remover. On the outside surface, clean it down to a perfectly smooth finish, ending the job by using crocus cloth and water so there will be no scratches.

Go over the entire body area using fine sandpaper and crocus cloth. If there are many layers of paint, it will be necessary to use a body grinder to remove the excessive or uneven layers of paint.

Clean the underside of each fender and apply two coats of rust-proof black enamel.

Apply two coats of rust-proof enamel to the inside areas of the hood. This will be the same color as the outside of the hood.

Use masking tape and cover the inside of the hood louvres.

Have the body shop do any necessary repairs to the body, to the fenders, and have the body sills welded together if they were restored previously.

After the body work has been finished, it is possible to save some money by masking off the areas which are not to be painted. Discuss this with the body man before proceeding with the painting.

Use masking tape and newspapers and cover over the areas which are not going to be painted as follows:

1. Mask over the strip between the cowl and rear of the hood.

2. Use newspapers to mask over the radiator and the shell.

3. Mask over the headlights and their flexible conduits.

4. Use newspapers and mask over the windshield and all window glass.

5. Mask over the tail light glass.

6. Mask over the bumpers (or remove them).

7. Mask the roof area down to the edge of the metal body.

8. Mask all door and rumble seat handles.

9. Cover the tires.

10. Remove the running boards.

14.3 Colors and combinations.

Of great importance is the striping and, fortunately, it is no longer a lost art. Nearly every wayside village has a hot rod enthusiast who has mastered the tapered brush. A careful perusal of the good-sized photographs reproduced in the MODEL A ALBUM will reveal the striping details of every model. The following color schemes for the 1929 model and others will indicate which striping colors were used with each tone grouping. On the following pages are IM-numbers which can be translated into color formulae. Between the two and the Sherwin-Williams lacquer mixes, it should be possible to come up with a true duplicate of the car's finish.

COLORS - MODEL A

1928

PHAETON, TUDOR SEDAN, ROADSTER, COUPE, SPORT COUPE: Niagara Blue (dark or light) body with French Gray belt, reveals, and stripe. Arabian Sand (dark or light) body with French Gray belt, reveals and stripe. Dawn Gray (dark or light) body with French Gray belt, reveals and stripe. Gun Metal Blue body with French Gray belt, reveals and stripe. On open cars the molding carries the same colors that are used on the reveals of closed cars, the stripe also being added.

TUDOR SEDAN (adopted February, 1928): Niagara Blue (light) body with Niagara Blue (dark) upper back, belt molding, and reveals, and French Gray stripe. Arabian Sand (dark) body with Copra Drab upper back, belt molding, and reveals, and French Gray stripe. Dawn Gray (dark) body with Gun Metal Blue upper back, belt molding and reveals, and French Gray stripe. Niagara Blue (dark) body with Niagara Blue (light) upper back, belt molding, and reveals, and French Gray stripe. Gun Metal Blue body with Black upper back, belt molding, and reveals, and French Gray stripe.

FORDOR SEDAN (production begun April 27, 1928): Balsam Green lower body, Pembroke Gray reveals, Valley Green upper molding and belt, and Old Ivory stripe. Copra Drab lower body, Copra Drab reveals, Seal Brown upper molding and belt, and French Gray Stripe.

FORDOR SEDAN: Copra Drab discontinued August 3, 1928. Replaced by Rose Beige for body, windshield and window reveals, with Seal Brown moldings and Orange stripe. Andalusite Blue for body, windshield and window reveals, with Arabian Sand (dark) molding and Orange stripe.

1929

PHAETONS AND ROADSTERS: Bonnie Gray Lower with Chelsea Blue Mouldings and Straw Color Stripe. Rose Beige Lower with Seal Brown Mouldings and Orange Stripe. Balsam Green Lower with Valley Green Mouldings and Cream Color Stripe. Andalusite Blue Lower with Black Mouldings and French Gray Stripe.

COUPES: Bonnie Gray Lower with Chelsea Blue Moulding and Reveals and Straw Stripe. Vagabond Green Lower with Rockmoss Green Moulding and Reveals and Straw Stripe. Rose Beige Lower with Seal Brown Moulding and Reveals and Orange Stripe. Andalusite Blue Lower with Black Moulding and Upper with Niagara Blue Light Reveals and French Gray Stripe.

TUDORS: Bonnie Gray Lower with Chelsea Blue Moulding, Reveals and Upper Back with Straw Stripe. Vagabond Green Lower with Rockmoss Green Moulding, Reveals and Upper Back with Straw Stripe. Rose Beige Lower with Seal Brown Moulding,. Reveals and Upper Back with Orange Stripe. Andalusite Blue Lower with Black Moulding and Upper Back, Niagara Blue Light Reveals with French Gray Stripe.

FORDORS: Bonnie Gray Lower with Chelsea Blue Belt and Upper Body with Bonnie Gray Reveals and Straw Stripe. Vagabond Green Lower with Rockmoss Green Belt and Upper Body with Vagabond Green Reveals and Straw Stripe. Bramble Brown Lower with Thorne Brown Belt and Upper Body with Bramble Brown Reveals and Neenah Cream Stripe. Rose Beige

Lower with Seal Brown Belt and Upper Body with Rose Beige Reveals and Orange Stripe. Andalusite Blue Lower with Andalusite Blue Belt and Upper Body with Niagara Blue Light Reveals and French Gray Stripe.

1930

TOWN SEDAN: Black, Chicle Drab and Copra Drab, Maroon, Kewanee Green.

CABRIOLET: Yellow and Seal Brown, Moleskin, Andalusite Blue, Kewanee Green, Black.

TUDOR SEDAN, FORDOR SEDAN (THREE WINDOW), FORDOR SEDAN DELUXE (TWO WINDOW), COUPE, SPORT COUPE, DELUXE COUPE: Andalusite Blue, Kewanee Green, Thorne Brown, Black, Chicle and Copra Drab.

PHAETON AND ROADSTER: Andalusite Blue, Kewanee Green, Thorne Brown, Chicle and Copra Drab, Black.

DELUXE PHAETON AND DELUXE ROADSTER: George Washington Blue with Tacoma Cream stripe and wheels. Raven Black with Aurora Red stripe and wheels. Stone Brown body and hood, Stone Gray (deep) molding with Tacoma Cream stripe and wheels (after 8-8-30). Brewster Green (medium) body and hood, Black molding with Apple Green stripe and wheels (after 9-25-30). Lombard Blue body, Hessian Blue stripe and wheels (after 10-3-30).

DELUXE COUPE, DELUXE SEDAN, TOWN SEDAN, AND CABRIOLET (after 9-25-30): Brewster Green (medium) hood, lower body, and window reveals with Black used for belt and above belt, Apple Green stripe and Black wheels.

1931

TOWN SEDAN, DELUXE SEDAN, VICTORIA SEDAN, DELUXE COUPE, DELUXE TUDOR SEDAN: Black upper body with Ford Maroon lower body and English Coach Vermilion stripe. Black upper body with Brewster Green lower body and Apple Green stripe. Copra Drab upper body with Chicle Drab lower body and Straw stripe. Elkpointe Green upper body with Kewanee Green lower body and Apple Green stripe. Black upper body with Black lower body and Apple Green (deep) stripe.

STANDARD FORDOR SEDAN, TUDOR SEDAN, COUPE: Black upper body with Thorne Brown lower body and Straw stripe. Black upper body with Lombard Blue lower body and Hessian Blue stripe. Copra Drab upper body with Chicle Drab lower body and Straw stripe. Elkpointe Green upper body with

112

Kewanee Green lower body and Apple Green stripe. Black upper body with Black lower body and Apple Green (deep) stripe.

CONVERTIBLE SEDAN (released for production May 22, 1931): Copra Drab body, hood, and moldings with Chicle Drab reveals, Straw stripe, and Tacoma Cream wheels. Washington Blue body, hood, and mouldings with Riviera Blue reveals and Tacoma Cream stripe and wheels. Brewster Green (medium) body, hood, and moldings with Tampa Red reveals, Vermilion stripe and Aurora Red wheels.

PHAETON AND ROADSTER: Thorne Brown upper body with Thorne Brown lower body and Straw stripe. Lombard Blue upper body with Lombard Blue lower body and Hessian Blue stripe. Copra Drab upper body with Chicle Drab lower body and Straw stripe. Elkpointe Green upper body with Kewanee Green lower body and Apple Green stripe. Black upper body with Black lower body and Apple Green (deep) stripe.

DELUXE PHAETON AND DELUXE ROADSTER: Riviera Blue upper body with Washington Blue lower body and Tacoma Cream stripe. Stone Gray (deep) upper body with Stone Brown lower body and Tacoma Cream stripe. Black upper body with Brewster Green lower body and Apple Green stripe. Black upper body with Black lower body and Apple Green (deep) stripe.

CABRIOLET: Black upper body with Brewster Green lower body and Apple Green stripe. Seal Brown upper body with Bronson Yellow lower body and Orange stripe. Moleskin Brown upper body with Light lower body and French Gray stripe. Lombard Blue upper body with Lombard Blue lower body and Hessian Blue stripe. Elkpointe Green upper body with Kewanee Green lower body and Apple Green stripe. Black upper body with Black lower body and Apple Green (deep) stripe. Ford Maroon body and deck with Black reveals and English Coach Vermilion stripe.

NOTE: After June 19, 1931 only three stripe colors were used: Apple Green, Red, or Tacoma Cream.

FORD MODEL A COLOR EQUIVALENTS

Color equivalents are based on DuPont finishes available at DuPont refinish distributors throughout the United States. Dulux finishes are air dry enamels while Duco finishes are lacquers. The original Model A's were finished in pyroxylin lacquers except for the fenders which were finished in black dipping enamel.

*Available in metallic quality only.

	Upper	Lower
Tudor and Coupe	Chelsea Blue (IM-120)	Bonnie Gray (IM-116)
	Rock Moss Green (IM-117)	Vagabond Green (IM-122)
	Seal Brown (IM-118)	Rose Beige (IM-119)
	Black	Andalusite Blue (IM-121)
Fordor	Chelsea Blue (IM-120)	Bonnie Gray (IM-116)
	Rock Moss Green (IM-117)	Vagabond Green (IM-122)
	Seal Brown (IM-118)	Rose Beige (IM-119)
	Adalusite Blue (IM-121)	Andalusite Blue
Cabriolet	Seal Brown (IM-118)	Cigarette Cream (IM-451)
Phaeton and Roadster	Bonnie Gray (IM-116)	Bonnie Gray
	Rose Beige (IM-119)	Rose Beige
	Andalusite Blue (IM-121)	Andalusite Blue
	Balsam Green (IM-124)	Balsam Green
Taxicab	Medium Cream (IM-125)	Duchess Blue (IM-123)
	Medium Cream (IM-125)	Balsam Green (IM-124)
Town Car	Black	Brewster Green (IM-1017)
	Black	Mulberry Maroon (IM-1046)
	Black	Thorne Brown (IM-283)
Town Sedan	Rock Moss Green (1M-117)	Vagabond Green (IM-122)
	Rock Moss Green (1M-117)	Lawn Green (IM-159)
Commercial Jobs	Rock Moss Green (1M-117)	Rock Moss Green (IM-117)

	Upper	Lower
Tudor Sedan	Chicle Drab (IM-91)	Copra Drab (IM-440)
	Kewanee Green (IM-546)	Elk Point Green (IM-543)
	Black	Andalusite Blue (IM-121)
	Thorne Brown (IM-283)	Thorne Brown
Two and Three Window Fordor Sedan	Thorne Brown (IM-283)	Thorne Brown
	Chicle Drab (IM-91)	Copra Drab (IM-446)
Sedan	Kewanee Green (IM-546)	Elk Point Green (IM-543)
	Black	Ford Maroon (IM-1011)
	Andalusite Blue (IM-121)	Andalusite Blue
	Black	Ford Maroon (IM-1011)
	Chicle Drab ((IM-91)	Copra Drab (IM-440)
Phaeton and Roadster	Thorne Brown (IM-283)	Thorne Brown
	Kewanee Green (IM-546)	Elk Point Green (IM-543)
	Chicle Drab (IM-91)	Copra Drab (IM-440)
Sport Coupe	Andalusite Blue (IM-121)	Andalusite Blue
	Kewanee Green (IM-546)	Elk Point Green (IM-543)
	Black	Andalusite Blue (IM-121)
	Chicle Drab (IM-283)	Copra Drab (IM-440)
	Thorne Brown (IM-283)	Thorne Brown
Standard Coupe	Andalusite Blue (IM-121)	Andalusite Blue
	Kewanee Green (IM-546)	Elk Point Green (IM-543)
	Chicle Drab (IM-91)	Copra Drab (IM-440)
	Thorne Brown	Thorne Brown
	Andalusite Blue (IM-121)	Andalusite Blue
	Seal Brown (IM-118)	Bronson Yellow

	Upper	Lower
Convertible Cabriolet	Moleskin Brown Lt. (IM-544) Kewanee Green (IM-546)	Elk Point Green (IM-543)
Tudor and Standard Fordor	Black	Lombard Blue (IM-1009)
Standard Coupe and Sport Coupe	Thorne Brown (IM-283) Elk Point Green (IM-543) Cobra Drab (IM-440)	Thorne Brown Chicle Drab (IM-91) Chicle Drab
DeLuxe Cabriolet	Elk Point Green (IM-543)	Kewanee Green (IM-546)
2-Window Fordor DeLuxe	Cobra Drab (IM-440)	Chicle Drab (IM-91)
Coupe, DeLuxe Sedan,Town	Black	Brewster Green Medium (IM-1017)
Sedan, Victoria Coupe	Black	Ford Maroon (IM-1011)
DeLuxe Phaeton	Black	Brewster Green Medium (IM-1017)
DeLuxe Roadster	Moulding—Stone Deep Gray (IM-1015) Moulding—Riviera Blue (IM-1013)	Stone Brown (IM-1016) Washington Blue Medium (IM-1014)
Phaeton and Roadster	Black Moulding—Elk Point Green (1M-543) Moulding—Cobra Drab (IM-440)	Lombard Blue (IM-1009) Thorne Brown (IM-283) Kewanee Green (IM-546) Chicle Drab (IM-91)
Cabriolet	Seal Brown (IM-118) Lombard Blue (IM-544)	Bronson Yellow (IM-545) Lombard Blue Moleskin Brown
Commercial Jobs	Blue Rock Green (IM-1012)	Blue Rock Green

The above (IM Numbers) can be converted by any Ditzler paint dealer into a formula that the desired colors may be made from.

14.5 Model A color equivalents.

MUNSELL CODE	COLOR	"DUCO"	"DULUX"
5 YR 5/11	Yukon Yellow		93-003
5 YR 6/13	Pegex Orange		93-1021
5 Y 8/5	Medium Cream	725	246-81979
5 Y 7/6	Cream	1559	246-57336
10 BGB 1/3	Blue Rock Green	2204-H*	2204-H*
5 G 2/2	Rock Moss Green	923-G	246-81572-G
5 G 3/4	Balsam Green		93-1855
5 BG 2/4	Vagabond Green		93-81872
5 Y 5/2	Arabian Sand Light		246-31470
10 YGY 3/2	Commercial Drab		93-35648
10 YRY 5/3	Pembroke Gray		None
10 GBG 4/2	Dawn Gray Light		None
5 G 3/1	Bonnie Gray	1293	246-57107
5 R 1/10	Rubelite Red	1497-M	1497-M
5 R 3/14	Vermilion		93-24119
5 PB 00/2	Lombard Blue		246-55106
10 BGB 1/2	Niagara Blue Dark		246-35961
10 BGB 2/4	Niagara Blue Light	783	246-55446
5 BG 2/2	Gun Metal Blue		93-81872
10 BGB 3/4	Duchess Blue	1345-G	246-57118-G
10 YRY 2/2	Mountain Brown		93-3836
10 GBG 2/4	Valley Green		246-34918
10 GYG 2/2	Highland Green		246-34116
10 GYG 3/1	Kewanee Green	785	246-71075

10 GYG 2/1	Elkspointe Green		246-35859
10 GBG 2/3	L'Anse Green Dark		93-6621
5 G 4/3	Lawn Green	662-G	246-62201-G
5 YR 3/5	Phoenix Brown		93-81412
10 YRY 5/4	Manila Brown	837*	202-55505*
5 YR 0/1	Thorne Brown		246-30340
10 YRY 3/1	Copra Drab	884	246-55551
5 Y 3/2	Chicle Drab		None
5 Y 4/2	Arabian Sand Dark		None
10 BGB 2/3	Washington Blue		246-34760
Neutral 1	Stone Gray Deep		246-51252
5 BG 3/2	Dawn Gray Dark	602	246-55134
10 YRY 4/3	Stone Brown		246-35922
5 GY 1/2	Brewster Green		246-54723
10 YRY 0/2	Seal Brown	658	246-60371
5 Y 7/7	Bronson Yellow		None
5 Y 2/1	Moleskin Brown Light		93-6846
5 R 0/3	Ford Maroon		None
10 BPB 00/2	Lombard Blue	914	246-81580
5 GY 3/3	Kewanee Green		93-546
5 YR 4/3	No. 1 Cord and		246-81467
	No. 3 Striped Cloth—Light		
5 YR 3/1	No. 2 Mohair and		246-50964
	No. 3 Striped Cloth—Dark		
5 YR 3/2	No. 4 Mohair		246-81467
10 YRY 4/3	No. 5 Leather, two toned		None
	textured: light area		
10 YRY 0/1	No. 5 Leather, two toned		None
	textured: dark area		
10 YRY 1/3	No. 6 Leather		93-3836

Sherwin-Williams Color Formulae for the following automobile lacquers (stated by volume):

RIVIERA BLUE: 41 parts Bone Black (Opex No. 31111), 33 parts Prussian Blue (Opex No. 31044), 24 parts Auto White (Opex No. 31001), 2 parts Toning Yellow (Opex No. 31163).

WASHINGTON BLUE: 61 parts Prussian Blue (Opex No. 31044), 30 parts Bone Black (Opex No. 31111), 9 parts Auto White (Opex No. 31001), Touch of Toning Yellow (Opex No. 31163).

TACOMA CREAM: 88 parts Auto White (Opex No. 31001), 12 parts Toning Yellow (Opex No. 31163), Touch of Red Oxide and Touch of Ultramarine Blue.

Tacoma Cream was used on the wheels, and a double pin stripe in Tacoma Cream ran over the Riviera Blue moulding reveals.

Although these mixes are for lacquer, they can be used for enamel by substituting comparable shades of mixing colors in enamel material.

The 1931 Model A Ford DeLuxe Roadster in the original factory combinations of Riviera Blue, Washington Blue, Tacoma Cream, and Black, was painted as follows:

Washington Blue was used on the hood and body, doors, rumble seat deck lid, etc.

Riviera Blue was used on the moulding reveals along the side of the hood, along tops of doors, and around the body. Two moulding reveals sweep down the rear of the body and along side the rumble seat deck lid.

Black was used on the fenders and splash aprons, bumper brackets, trunk carrier, etc.

CHAPTER 15

MAINTENANCE OF THE MECHANICAL BRAKE SYSTEM

15.1 General.

Correct operation of the brakes will require a periodical routine check of the components which comprise the braking system followed by the replacement, repair, or readjustment of any parts which may not be functioning correctly. The major problems are caused by wear and if any part is excessively worn it should be replaced.

The primary offenders to having good brakes are the brake shoes and the wheel drums. Most of the remaining offenders are due to sloppy mechanical actions caused by wear.

15.2 Brake shoes and brake drums.

1. Remove each wheel and wheel drum. Examine the brake shoes for uneven wear and excessive wear as evidenced by only part of the shoes being worn down and by showing of rivets in the shoes.

2. If the shoes are considerably worn, replace them.

3. If the brake drums are scored (gouged by the shoe rivets), take them to a brake shop for advice. If they can be turned down without the removal of a large amount of metal (the machinist knows the limit), have them turned down. If they have already been turned down several times (the machinist knows original diameter), replace the drums with new ones.

Don't be misled into thinking that wheel drums can always be turned down one more time. There is a limit at which point even oversized brake shoes will no longer reach the drums.

Turning down the wheel drums, if allowable, and replacing the brake shoes with oversize ones will eliminate many thousandths of inches of play.

Use a good grade of rust-proof black paint on the outside of the drums and on the backing plates.

15.3 Brake springs.

Don't fool around with the brake return springs. Buy a Model A brake spring kit and replace all of them.

15.4 Roller track.

The roller track may be excessively worn. If a new one cannot be located, remove the rivets that secure it to the backing plate. Build it up with weld and grind it to a straight surface.

15.5 Brake adjusting shaft pin.

If the operating shafts are worn, replace them, or build them up with weld (only if absolutely necessary). The shoe roller pins, brake shoe rollers, and rear brake cams should all be replaced to eliminate slack or slop.

15.6 Service brake cross shaft.

Remove the brake rods from the cross shaft and examine the cross shaft for slop and runout. Build up the ends of the cross shaft and their holders, or replace them.

15.7 Brake rods, clevises, and clevis pins.

If the brake rods are bent, replace them. If the holes in any clevis are enlarged or elongated, replace the clevis. If the clevis pins are worn or grooved, replace them.

15.8 Brake pedal shaft and bushing.

Examine the brake pedal and the shaft for excessive wear as evidenced by slop after removing the pedal to cross shaft rod. If the shaft is worn, replace it. If the bushing is worn, drive it out and replace it.

Check the clutch pedal in the same manner at this time.

Replace all the cotter pins.

15.9 Adjustment.

Each front wheel brake is adjusted by a wedge at the top of its backing plate. Each rear wheel brake is adjusted by a wedge at the rear of its backing plate. Turning the lugs (or wedges) clockwise will tighten the brake, counterclockwise will loosen the brake.

The rear brakes should be adjusted so that they start to hold when the brakes are applied with the pedal being depressed one inch.

When the brake pedal reaches $1^1/_2$ inches, the rear brakes should tighten but not lock. The front brakes, at this point, should be adjusted until they start to hold.

With the brake pedal at two inches, the rear wheels should lock, and the front wheels should be very tight, but not locked.

During a road test, apply full pressure to the brake pedal. The rear wheels should slide and the front wheels should make a heavy print. The two front wheels should work evenly and the two rear wheels should work evenly.

If the king pins are badly worn, it will be dangerous to operate the brakes as they may bind. If this is the case, it will be necessary to replace the king pins.

CHAPTER 16

OPERATION AND MAINTENANCE

16.1 Liquid capacities.

Fill the radiator and cooling system with clean fresh water; approximately three gallons. Keep it full at all times.

The gasoline tank capacity is ten gallons. Be sure that the small vent hole in the gasoline tank cap is kept open.

Oil pan capacity is five quarts. Keep it full at all times. When inserting the oil dip stick, make certain that both ends go into the opening.

16.2 Starting the engine.

1. Put the gear shift in its neutral position.

2. Advance the throttle lever about three notches or until the accelerator pedal just starts to move downwards.

NOTE: Advance means to move the throttle (or spark) lever from its upper maximum position towards a lower position. Retard means to move the throttle (or spark) lever from a low position towards a higher position.

3. Place the spark lever in its retarded (up) position.

CAUTION

Always place the spark lever in its retarded position when starting the engine. This will prevent the engine from kicking back and damaging the starter.

4. Turn the ignition key on.

5. If the engine is cold, turn the carburetor adjusting rod one full turn to the left. If the engine is warm, leave the carburetor adjusting rod in its normal operating position, one quarter turn open.

6. Pull the choke rod out, step on the starter button, push the choke rod down and advance the spark lever. When the engine has warmed up, turn the carburetor adjusting rod back to its normal operating position, one quarter turn open. Do not use the choke rod excessively as engine flooding will develop. If the engine should become flooded, advance the throttle lever, push the choke rod in and turn the engine over several times to exhaust the gas.

7. If starting the engine for the first time, it will be necessary to make new settings for the carburetor adjustments. Turn on the gas shutoff valve. Turn the idler adjustment screw $1^1/_2$ to $3^1/_2$ turns counterclockwise. Start the engine and turn the idler screw for the smooothest operation of the engine. The other adjust controls the idling speed of the engine and should be

set at the position which allows the engine to idle at a slow speed just above stalling speed.

16.3 Driving the car, lubrication.

1. Keep the spark lever fully advanced during all normal driving. Whenever the engine is under a heavy load, the spark lever should be retarded just enough to prevent a spark knock.

2. With a new or rebuilt engine, the car should not be driven more than 30-35 miles per hour during the first 500 miles of operation.

3. Change the oil at 500 miles.

4. Maintain the tire air pressure at 35 pounds.

5. Keep the battery filled with distilled water.

6. Do not drive the car with one foot on the clutch pedal as this will cause premature clutch wear.

7. The clutch pedal adjustment should be maintained at $3/4$-inches of free play.

8. Replace the oil every 500 miles and change it only when the engine is warm.

9. Drain the differential housing every 5,000 miles, flush it out with kerosene and refill it with 600W lubricant.

10. Drain the transmission housing every 5,000 miles, flush it out with kerosene and refill it with 600W lubricant.

11. Remove the hand hole cover from the clutch housing, turn the clutch thrust bearing until the grease fitting is at the top and grease it.

12. If adjustment of the fan belt becomes necessary, tighten the generator just enough to prevent the belt from slipping.

13. If the water pump shaft leaks water, tighten the packing nut with a screwdriver just enough to stop the leak.

14. Flush out the cooling system several times each year.

15. Adjust the breaker points (when necessary) as follows:
Take off the distributor cap, rotor and body. Turn the engine over with its crank until the breaker arm rests on one of the four high points of the cam with the breaker points fully opened. Loosen the lock screw and turn the contact screw (see Figure 6-8) until the gap is .015 to .018 inches.

16. To time the ignition, see "Timing the engine," paragraph 6.15 in Chapter 6.

17. During the winter months the generator charging rate should be 14 amperes. During the summer months the generator charging rates should be 10 amperes. Adjust the generator charging rate as follows:
Remove the generator cover and loosen the field brush holder

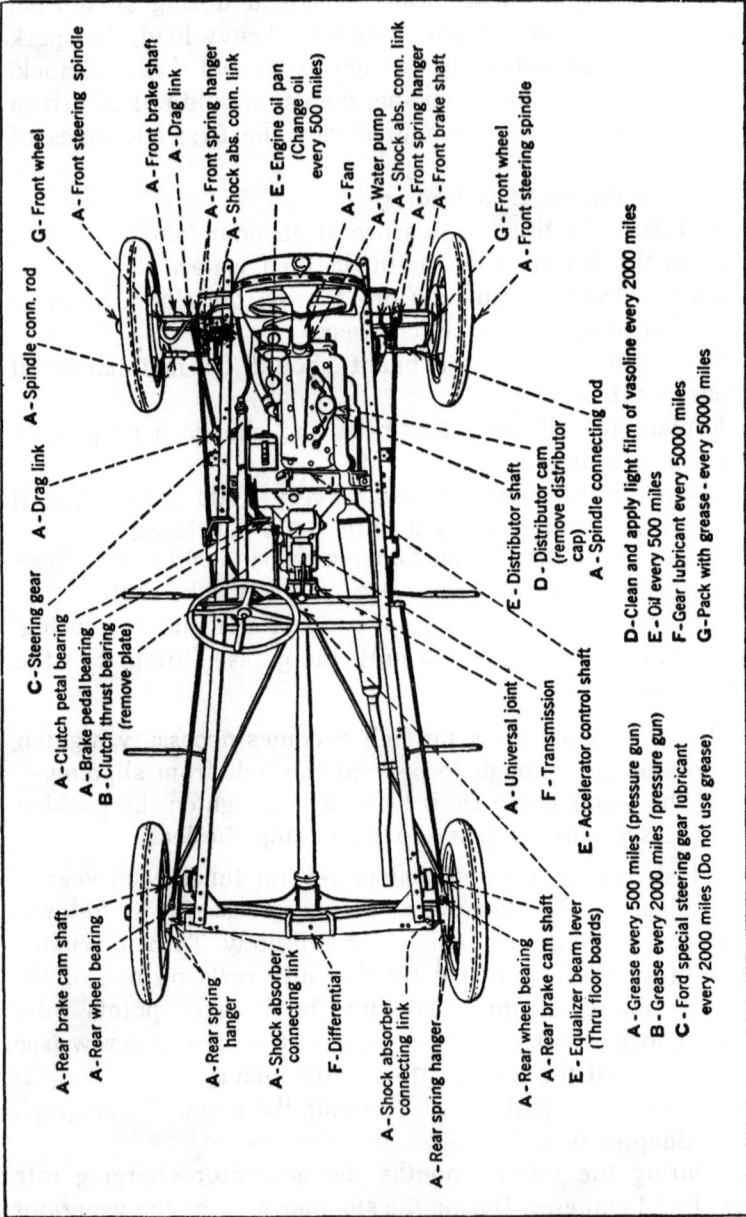

A - Clutch petal bearing
A - Brake pedal bearing
B - Clutch thrust bearing
(remove plate)

C - Steering gear

A - Drag link

A - Spindle conn. rod

A - Front brake shaft
A - Drag link
A - Front spring hanger
A - Shock abs. conn. link

G - Front wheel

A - Front steering spindle

E - Engine oil pan
(Change oil
every 500 miles)

A - Fan

A - Water pump
A - Shock abs. conn. link
A - Front spring hanger
A - Front brake shaft

G - Front wheel

A - Front steering spindle

E - Distributor shaft

D - Distributor cam
(remove distributor
cap)

A - Spindle connecting rod

A - Universal joint

F - Transmission

E - Accelerator control shaft

A - Rear brake cam shaft
A - Rear wheel bearing

A - Rear spring
hanger

A - Shock absorber
connecting link

F - Differential

A - Shock absorber
connecting link

A - Rear spring hanger

A - Rear wheel bearing
A - Rear brake cam shaft

E - Equalizer beam lever
(Thru floor boards)

A - Grease every 500 miles (pressure gun)
B - Grease every 2000 miles (pressure gun)
C - Ford special steering gear lubricant
every 2000 miles (Do not use grease)

D - Clean and apply light film of vasoline every 2000 miles
E - Oil every 500 miles
F - Gear lubricant every 5000 miles
G - Pack with grease - every 5000 miles

Figure 16.1. Lubrication Chart

lock screw. This is the brush holder which operates in a slot and provided with a locking screw. Increase the charging rate by shifting the field brush holder in the direction of rotation. Decrease the charging rate by shifting the field brush holder in the opposite direction. The generator output is indicated by the ammeter on the instrument panel.

18. Keep the ignition coil and its connections clean at all times.

19. The correct spark plug gap is .035 inches.

20. Keep all steering gear nuts and belts tight by checking them every few weeks.

21. Make a periodical check of play in the front ends (spindle bolts and wheel bearings).

22. Adjust front wheel bearing (when necessary) as follows: Remove the wheel and cotter pin. Tighten the spindle nut until the drum just starts to bind. Back off on the adjusting nut one or two notches until the drum can rotate freely. Insert the cotter pin and remount the wheel.

23. Lubricate the springs occasionally with oil or graphite.

24. Fill the (original type) shock absorber reservoirs every 5,000-10,000 miles with commercial glycerine. NEVER USE OIL as a substitute.

25. Lubricate the flexible speedometer cable every 5,000 miles.

26. Have the speedometer lubricated every 10,000 miles by a specialist. Do not break the seal on the speedometer.

16.4 **Summary of engine troubles and their causes.**

If the starter turns engine over freely but engine will not run, check the following:

1. Ignition switch.

2. Gasoline tank is empty or supply is shut off.

3. If engine is cold, mixture may not be rich enough — choke rod not pulled back.

4. Warm engine — over-choking.

5. Breaker points too close. The correct adjustment is .015 to .018 inches.

6. Spark plugs gaps too wide. Correct gap is .035 inches.

7. Water in sediment bulb or carburetor.

16.4.1 **Starter fails to turn engine over.**

1. Battery run down. A quick way to check this is to turn on the lights, and depress the starter switch. If the battery is weak, the lights will go out or grow quite dim. If the battery is run down, have it recharged.

2. Loose or dirty battery connections — see that both the negative and positive battery terminals are clean and tight. These connections should be checked regularly.

16.4.2 Missing at low speed.

1. Gas mixture too rich or too lean.

2. Too close a gap between spark plug points. The correct gap is .035 inches.

3. Breaker points improperly adjusted, badly burnt or pitted.

4. Fouled spark plugs. Plugs should be occasionally cleaned and the gaps checked.

5. Water in gasoline.

16.4.3 Missing at high speed.

1. Insufficient gasoline flowing to carburetor due to gasoline line filter screen being partly clogged.

2. Gas mixture too rich or too lean.

3. Water in gasoline.

16.4.4 Engine stops suddenly.

1. Gasoline tank empty.

2. Dirt in fuel line or carburetor.

3. Gas mixture too lean.

16.4.5 Engine overheats.

1. Lack of water — radiator should be kept well-filled.

2. Lack of oil — check oil level.

3. Fan belt loose or slipping.

4. Carbon deposit on piston heads and in combustion chamber. This can be corrected by taking off the cylinder head and removing the carbon.

5. Incorrect spark timing.

6. Gas mixture too rich.

7. Water circulation retarded by sediment in radiator.

16.4.6 Engine knocks.

Carbon knock.

Carbon knock is caused by a deposit of carbon in the combustion chamber and on the piston heads. Take off the cylinder head and remove the carbon.

Ignition knock.

Ignition knock, under ordinary driving conditions, could be caused by the engine being out of time. If the engine is overheated, check the conditions listed above under "Engine overheats." If a bearing has become loose, it will be necessary to check the main bearings, the crankshaft, and the connecting rods.

Do not mistake an ignition knock for a loose bearing. Ignition knocks usually occur when the car is suddenly accelerated or when ascending a steep grade or traveling through heavy sand with the spark lever fully advanced. Slightly retarding the spark lever eliminates the knock. The spark should be advanced as soon as normal road conditions are encountered. Loose bearings knock all the time the engine is in use and regardless of spark timing.

CHAPTER 17

AUTHENTICITY

Throughout the entire content of this book the word "authentic" has been used often. Before a restoration can be considered as being complete and final, it is of utmost importance to have retained the actual authenticity of the car as it was when it was manufactured many years ago.

If the availability of money has been of no major problem, the final results will be the possession of a Model A Ford operating in like-new condition after all defective or missing parts replaced with new ones.

The original headlamps and mechanical brake system will have been retained and the roof material and upholstery will have been restored with the types of material that were originally used.

If the words "safety" or "safety factor" mean a lot to the restorer, then the headlamps will have been replaced with sealed beam units and the mechanical brakes replaced with hydraulic brakes.

The possession of a fine car with a first class exterior gleaming finish will allow the restorer to be justifiably proud of the workmanship applied to the restoration of the car and of the expenditure of money into an antique automobile which should receive admiration wherever it may be seen.

On the other hand, if the availability of money has been a major problem, the final results will be the possession of a Model A Ford operating in good condition but still needing more work to be performed on it before it can be considered as a true and complete authentic restoration. As time progresses, the restorer should continue to expend efforts into completing the restoration by replacing all repaired units with new ones. By so doing, the end result will be a truly authentic restoration, one of which the restorer can be justifiably proud.

NOTE: On the following pages will be found scale drawings with dimensions and body section illustrations which help in restoring the car which has badly deteriorated or has been butchered. Also included are some of the most prominent identification characteristics of the various year models. However, for minute details which will help you establish the date on which your car was made, see "How To Restore The Model A Ford" which has a chronological history of Model A production.

The Phaeton

1929

The Coupe

1929

128

1929

The Business Coupe

The Roadster

1929

130

The Tudor Sedan

1929

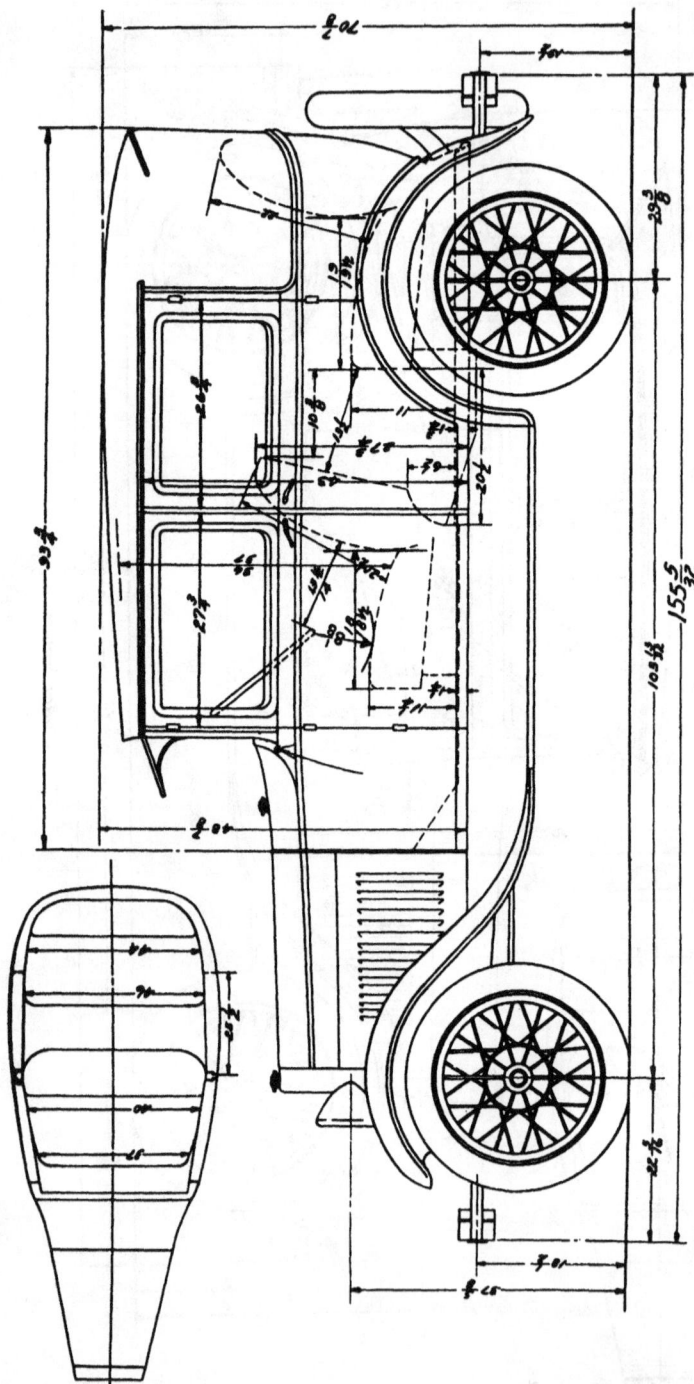

The Fordor Sedan

1929

The Cabriolet

1929

133

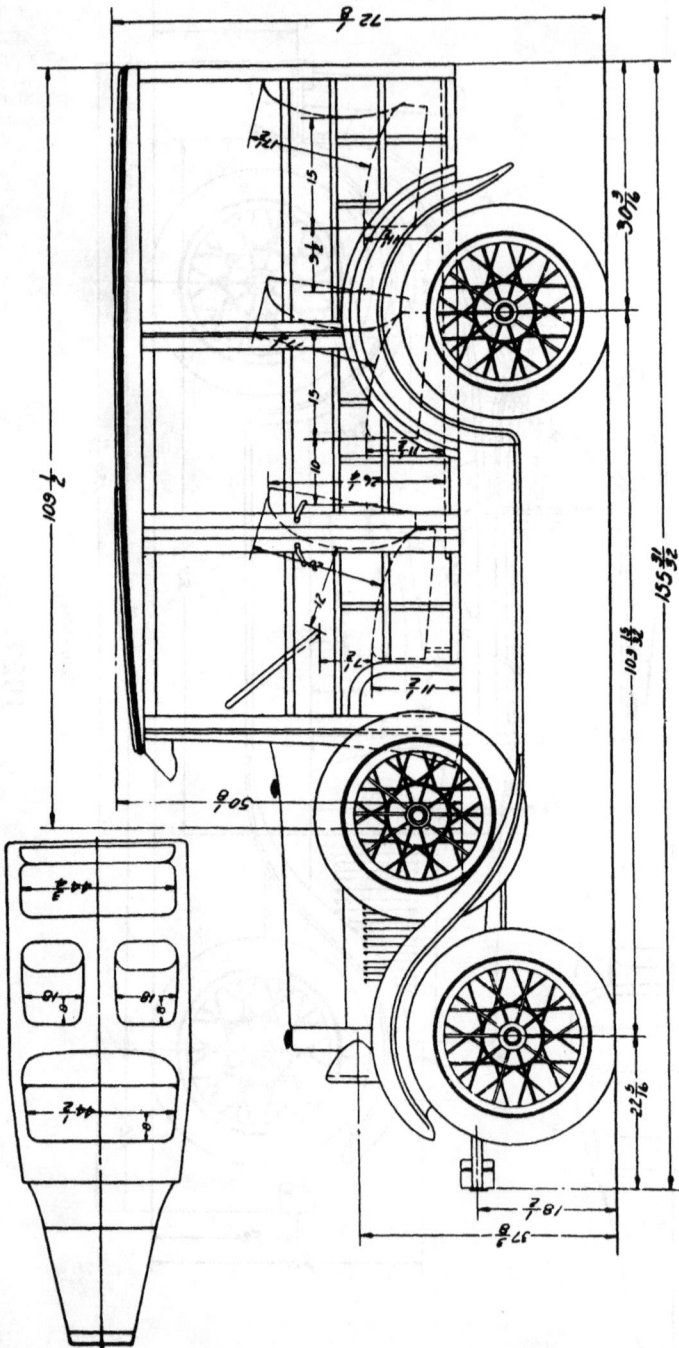

The Station Wagon
1929

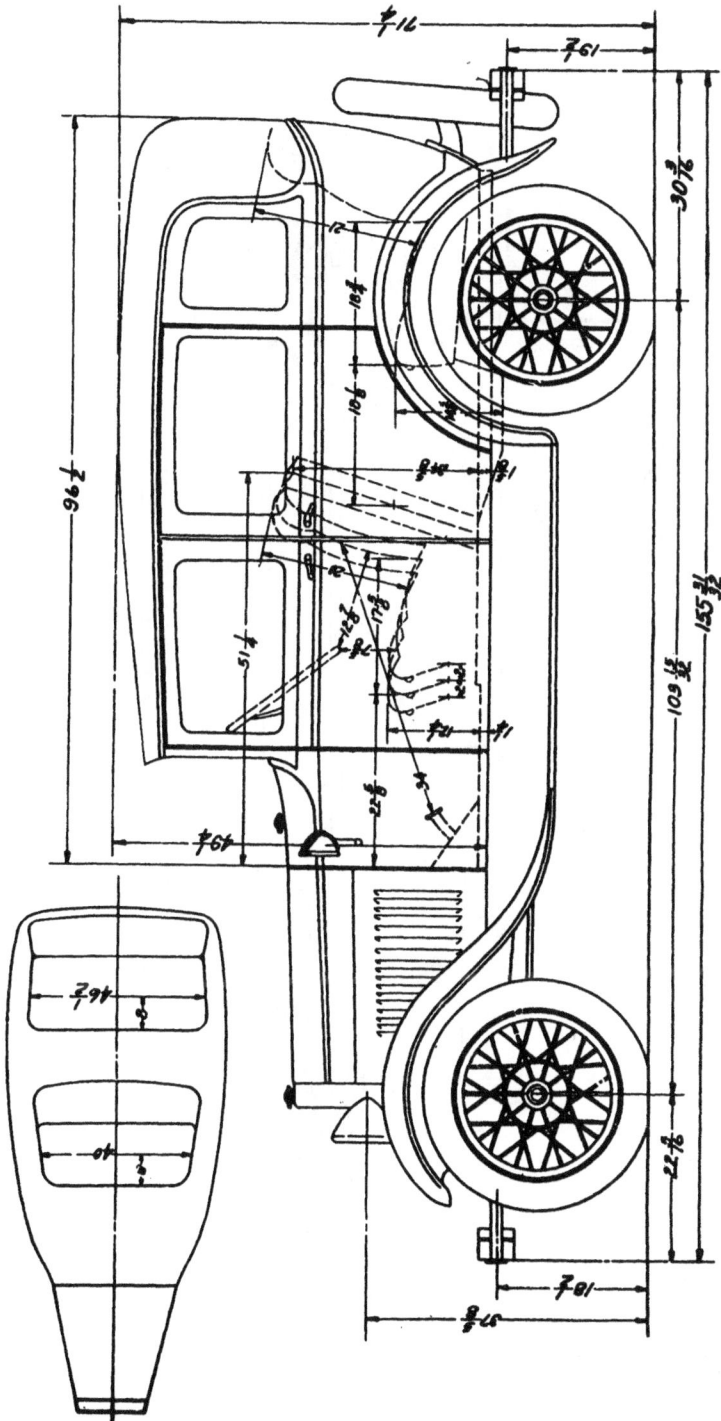

The Town Sedan

1929

135

The Town Car

1929

136

The Taxicab 1929

The Roadster

1930

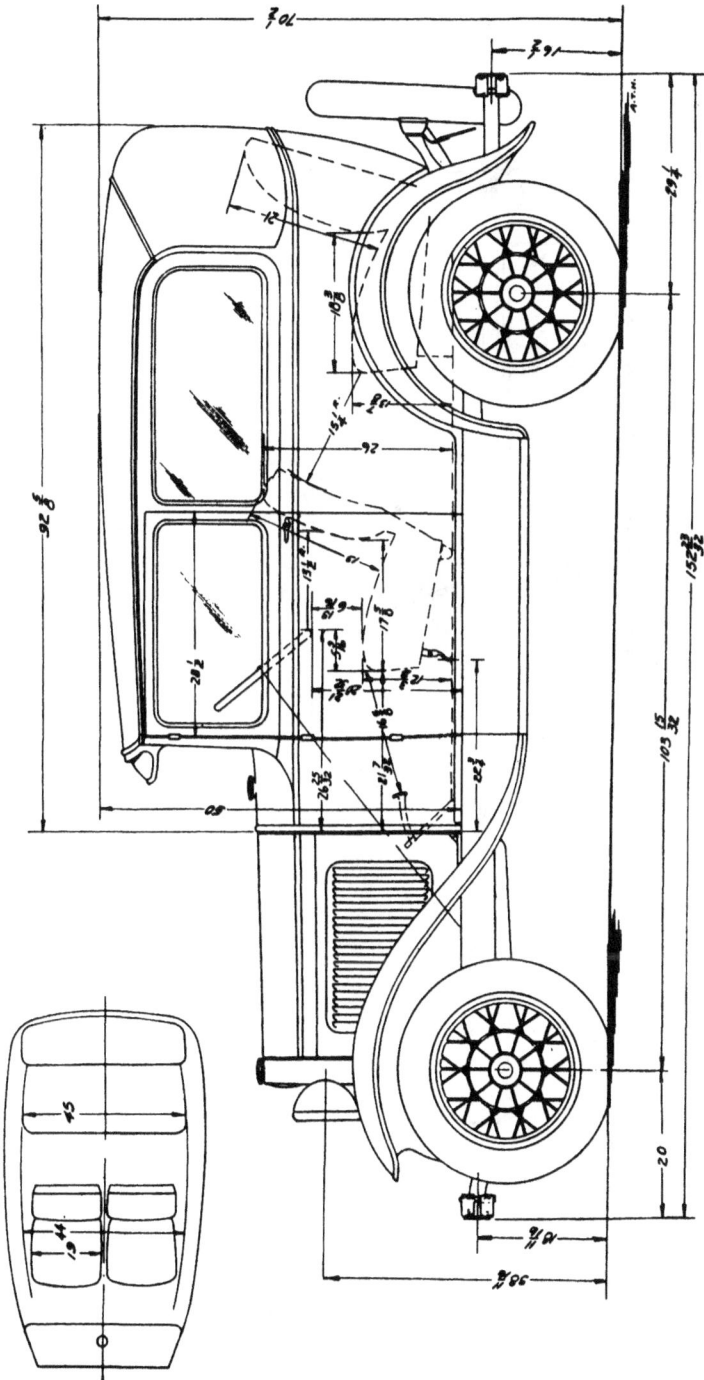

The Tudor Sedan 1930

139

Three-Window Fordor Sedan
(Dimensions same for De Luxe Sedan)

1930

The Sport Coupe

(Dimensions approximately same for Cabriolet)

1930

141

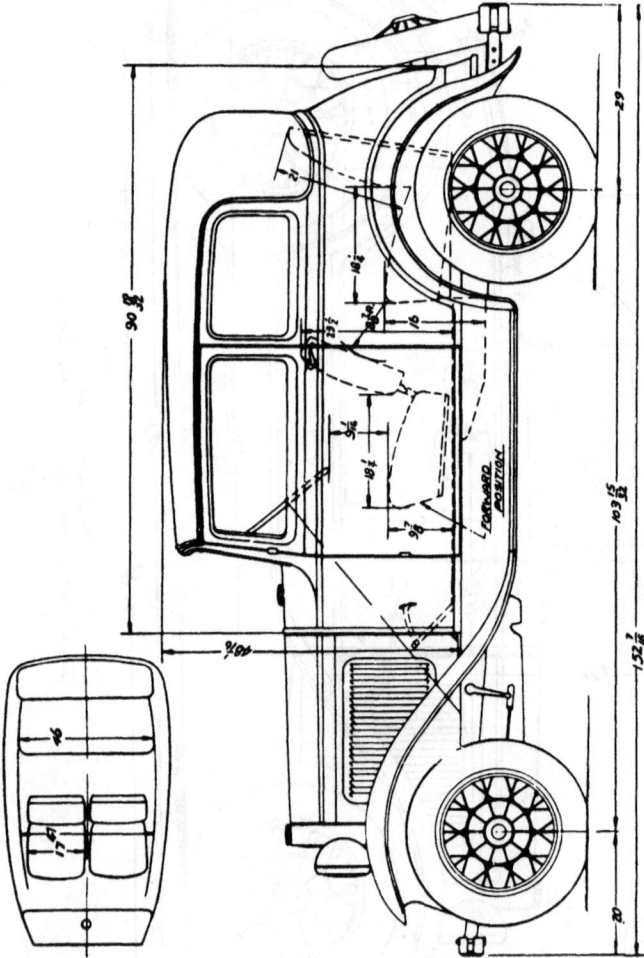

1931 VICTORIA

(Dimensions approximately same for De Luxe Phaeton, except it is
3 inches longer from back of front seat)

142

1931 CLOSED CAB PICK-UP

This dimensional drawing of the Closed Cab Pick-Up truck shows its construction on the standard MODEL A Ford chassis with the 103½" wheelbase.

143

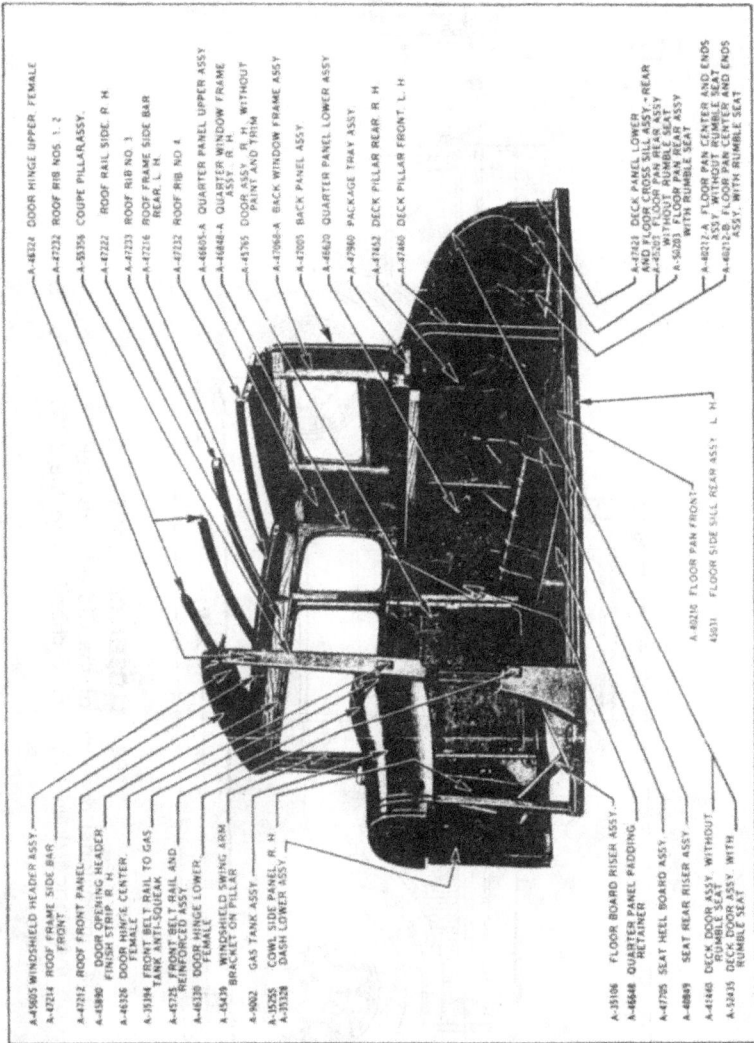

A-45605 WINDSHIELD HEADER ASSY
A-47214 ROOF FRAME SIDE BAR FRONT
A-47292 ROOF FRONT PANEL
A-45890 DOOR OPENING HEADER FINISH STRIP R. H
A-45326 DOOR HINGE CENTER FEMALE
A-35394 FRONT BELT RAIL TO GAS TANK ANTI-SQUEAK
A-45225 FRONT BELT RAIL AND REINFORCED ASSY
A-45130 DOOR HINGE LOWER FEMALE
A-45439 WINDSHIELD SWING ARM BRACKET ON PILLAR
A-9602 GAS TANK ASSY
A-15255 COWL SIDE PANEL R. H
A-35328 DASH LOWER ASSY

A-35106 FLOOR BOARD RISER ASSY.
A-46648 QUARTER PANEL PADDING RETAINER
A-47205 SEAT HEEL BOARD ASSY
A-46849 SEAT REAR RISER ASSY
A-47240 DECK DOOR ASSY WITHOUT RUMBLE SEAT
A-52435 DECK DOOR ASSY WITH RUMBLE SEAT

A-45224 DOOR HINGE UPPER FEMALE
A-47232 ROOF RIB NOS. 1, 2
A-55356 COUPE PILLAR ASSY.
A-47222 ROOF RAIL SIDE. R. H
A-47233 ROOF RIB NO 3
A-47216 ROOF FRAME SIDE BAR REAR. L. H
A-47232 ROOF RIB NO 4
A-46695-A QUARTER PANEL UPPER ASSY
A-46848-A QUARTER WINDOW FRAME ASSY R. H
A-47365 DOOR ASSY. R. H WITHOUT PAINT AND TRIM
A-47068-A BACK WINDOW FRAME ASSY
A-47005 QUARTER PANEL LOWER ASSY
A-46620 PACKAGE TRAY ASSY
A-47452 DECK PILLAR REAR R. H
A-47460 DECK PILLAR FRONT L. H

A-47421 DECK PANEL LOWER AND FLOOR CROSS SILL ASSY. - REAR
A-45201 FLOOR PAN REAR ASSY WITHOUT RUMBLE SEAT
A-56203 FLOOR PAN REAR ASSY WITH RUMBLE SEAT
A-46212-A FLOOR PAN CENTER AND ENDS ASSY. WITHOUT RUMBLE SEAT
A-46212-B FLOOR PAN CENTER AND ENDS ASSY. WITH RUMBLE SEAT

A-60210 FLOOR PAN FRONT
49031 FLOOR SIDE SILL REAR ASSY. L. H

144

A-79051 - BACK BELT RAIL CENTER
A-69674 - B-QUARTER LOCK PILLAR UPPER ASSY R.H.
A-70262 - ROOF TO LOCK PILLAR HINGES
A-70239 - B ROOF BOW NO. 1 ASSY
A-70271 B - ROOF LANDAU IRON TO QUARTER LOCK PILLAR UPPER BRKT L.H.
A-70281 - ROOF SIDE FOLDING HINGE ASSY L.H.
A-70440 - PACKAGE TRAY ASSY
A-70235 B - ROOF RAIL FRONT
A-68605 - WINDSHIELD HEADER ASSY
A-68314 - COUPE PILLAR ASSY L.H.
A-68725 - B FRONT BELT RAIL ASSY
A-46254 - C COWL PANEL ASSY
A-46131 - DASH INTERMEDIATE ASSY
A-46128 - DASH LOWER ASSY
A-68706 - DOOR ASSY L.H. LESS PAINT AND TRIM
A-74030 - COWL MOUNDING RETAINER
A-68904 - QUARTER PANEL ASSY L.H.
A-68508 - B FLOOR SIDE SILL REAR ASSY R.H.
A-68540 - FLOOR CROSS SILL FRONT
A-70705 - SEAT HEEL BOARD ASSY
A-63490 - SEAT TOOL BOX BOTTOM
A-68045 - FLOOR CROSS SILL CENTER

A-52531 - DECK DOOR BODY HINGE ASSY L.H.
A-70052 - BACK BELT RAIL CORNER ASSY R.H.
A-70272 - ROOF LANDAU IRON BRACKET R.H.
A-70468 - DECK PILLAR FRONT R.H.
A-41502-A DECK DRAIN TROUGH SIDE R.H.
A-70452 - DECK PILLAR REAR R.H.
A-52435 - DECK DOOR ASSY

A-68854 - FLOOR CROSS SILL REAR ASSY
A-47420 - A DECK PANEL LOWER ASSY
A-68203 - FLOOR PAN REAR ASSY
A-52486 - DECK DOOR BUMPER BRACKET
A-52527 - DECK DOOR HINGE L.H.
A-68212 - FLOOR PAN CENTER ASSY
A-52642 - DECK SIDE CARDBOARD RETAINER LOWER
A-68209 - FLOOR PAN FRONT ASSY
A-59690 - QUARTER LOCK PILLAR LOWER R.H.

Sectional View of Cabriolet (1929 design)

145

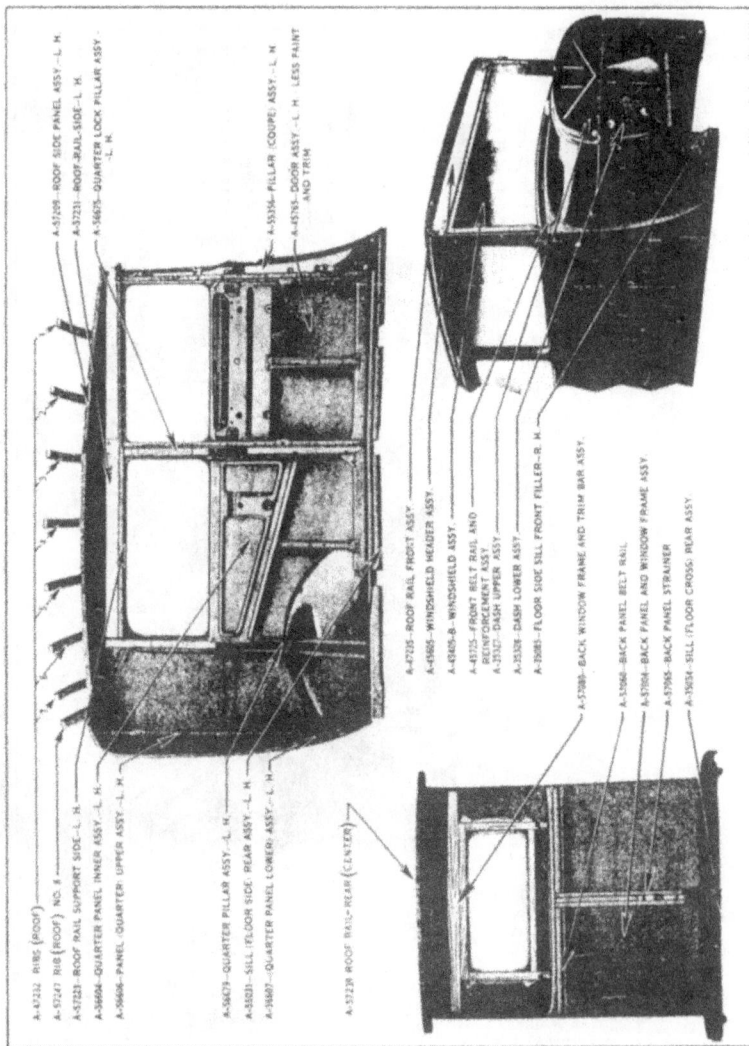

A-47282 RIBS (ROOF)
A-51247 RIB (ROOF) NO. 8
A-51223 ROOF RAIL SUPPORT SIDE—L. H.
A-56604 QUARTER PANEL INNER ASSY.—L. H.
A-56606 PANEL, QUARTER UPPER ASSY—L. H.

A-56679 QUARTER PILLAR ASSY—L. H.
A-58031 SILL (FLOOR SIDE REAR ASSY—L. H.
A-56607 QUARTER PANEL (LOWER) ASSY—L. H.

A-51228 ROOF RAIL—REAR (CENTER)

A-51209 ROOF SIDE PANEL ASSY—L. H.
A-57231 ROOF RAIL SIDE—L. H.
A-56625 QUARTER LOCK PILLAR ASSY—L. H.

A-55356 PILLAR (COUPE) ASSY—L. H. LESS PAINT AND TRIM
A-45785 DOOR ASSY—L. H.

A-47235 ROOF RAIL FRONT ASSY
A-45695 WINDSHIELD HEADER ASSY
A-45405-B WINDSHIELD ASSY
A-45725 FRONT BELT RAIL AND REINFORCEMENT ASSY
A-21507 DASH UPPER ASSY
A-21508 DASH LOWER ASSY
A-29085 FLOOR SIDE SILL FRONT FILLER—R. H.

A-57880 BACK WINDOW FRAME AND TRIM BAR ASSY
A-57060 BACK PANEL BELT RAIL
A-57804 BACK PANEL AND WINDOW FRAME ASSY
A-57805 BACK PANEL STRAINER
A-15654 SILL (FLOOR CROSS) REAR ASSY

—Sectional View of the Tudor

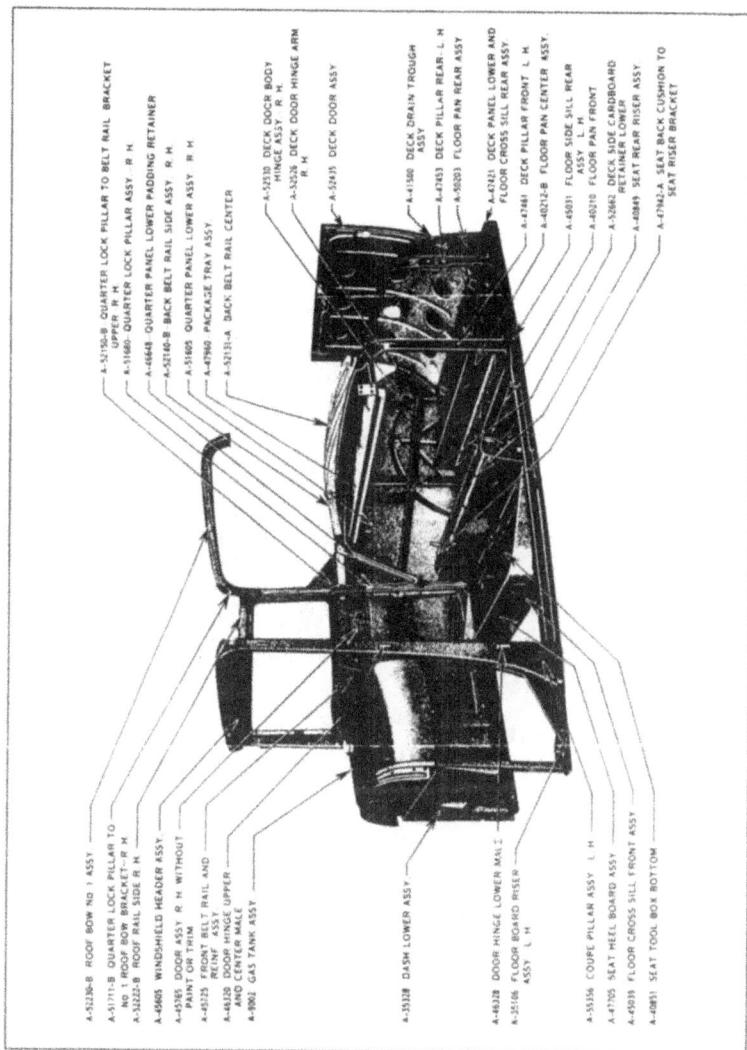

Sectional View of Sport Coupe

A-52230-B ROOF BOW NO 1 ASSY
A-51711-B QUARTER LOCK PILLAR TO NO 1 ROOF BOW BRACKET R H
A-51222-B ROOF RAIL SIDE R H
A-45605 WINDSHIELD HEADER ASSY
A-45765 DOOR ASSY R H WITHOUT PAINT OR TRIM
A-45125 FRONT BELT RAIL AND REINF ASSY
A-48320 DOOR HINGE UPPER AND CENTER MALE
A-9062 GAS TANK ASSY

A-35528 DASH LOWER ASSY
A-48628 DOOR HINGE LOWER MALE
R-25106 FLOOR BOARD RISER ASSY L H
A-55556 COUPE PILLAR ASSY L H
A-47105 SEAT HEEL BOARD ASSY
A-45039 FLOOR CROSS SILL FRONT ASSY
A-40851 SEAT TOOL BOX BOTTOM

A-52150-B QUARTER LOCK PILLAR TO BELT RAIL BRACKET UPPER R H
R-51680 QUARTER LOCK PILLAR ASSY R H
A-46648 QUARTER PANEL LOWER PADDING RETAINER
A-52140-B BACK BELT RAIL SIDE ASSY R H
A-51605 QUARTER PANEL LOWER ASSY R H
A-47960 PACKAGE TRAY ASSY
A-52151-A BACK BELT RAIL CENTER

A-52510 DECK DOOR BODY HINGE ASSY R H
A-52526 DECK DOOR HINGE ARM R H
A-52415 DECK DOOR ASSY
A-41560 DECK DRAIN TROUGH ASSY
A-47451 DECK PILLAR REAR L H
A-50203 FLOOR PAN REAR ASSY
A-47421 DECK PANEL LOWER AND FLOOR CROSS SILL REAR ASSY
A-47461 DECK PILLAR FRONT L H
A-40212-B FLOOR PAN CENTER ASSY
A-45031 FLOOR SIDE SILL REAR ASSY L H
A-40270 FLOOR PAN FRONT
A-52662 DECK SIDE CARDBOARD RETAINER LOWER
A-40849 SEAT REAR RISER ASSY
A-47942-A SEAT BACK CUSHION TO SEAT RISER BRACKET

A-156266—ROOF REAR BOW ASSY
A-156828—BODY LOCK PILLAR TO ROOF RAIL BRACE ASSY.
A-156290—ROOF NETTING

A-156282—ROOF RIB NO. 1
A-156283—ROOF RIB NO. 2
A-156284—ROOF RIB NO. 3
A-156285—ROOF RIB NO. 4
A-156286—ROOF RIB NO. 5
A-156287—ROOF RIB NO. 6

A-156860—REAR SEAT FRAME-REAR
A-156235—ROOF RAIL FRONT ASSY.
A-156228—ROOF RAIL SIDE INNER ASSY. R.H.
A-196228—ROOF RAIL SIDE AND BLOCK ASSY.—R.H.
A-60756—COUPE PILLAR ASSY.—L.H.
A-60364C. COUPE PILLAR COVER PANEL ASSY.
A-60354C. COWL PANEL ASSY.
A-35532 DASH UPPER ASSY.
A-6030—DASH INTERMEDIATE ASSY.
A-6028 DASH LOWER ASSY.

A-156896—BODY LOCK PILLAR AND PANEL ASSY—L.H.
A-156615—BODY LOCK PILLAR TO SILL BRACE
A-7276—FLOOR CROSS SILL FRONT FLOOR BOARD SUPPORT ASSY.
A-156040—FLOOR CROSS SILL-FRONT
A-157631—QUARTER HINGE PILLAR TO WHEELHOUSE BRACE—L.H.
A-150044—FLOOR CROSS SILL CENTER ASSY.

A-156232—ROOF RAIL SIDE FRONT STRAINER—R.H.
A-156290—ROOF DOME LIGHT BLOCK AND SPLINE ASSY.
A-156078—BACK WINDOW FRAME STRAINER
A-156234—ROOF RAIL SIDE REAR STRAINER—R.H.
A-18660—BACK BELT RAIL ASSY
A-156267—ROOF RAIL REAR ASSY.

A-156440—BODY LOCK PILLAR BUFFER CASING
A-157674—QUARTER PILLAR ASSY.—R.H.
A-157718—QUARTER BELT RAIL—R.H.
A-157690 QUARTER PILLAR BLOCK
A-156982 BACK PANEL SIDE STRAINER
A-156960 BACK PANEL CENTER STRAINER ASSY
A-157685—QUARTER STRAINER LOWER R.H.
A-18091—BACK PANEL TIRE CARRIER BRACKET SUPPORT-INNER
A-18090—BACK PANEL TIRE CARRIER BRACKET SUPPORT OUTER ASSY

A-156515—QUARTER PILLAR TO SILL BRACE
A-156054—FLOOR CROSS SILL REAR ASSY
A-18958—REAR SEAT FRAME SIDE
A-157654—QUARTER HINGE PILLAR TO SILL BRACE—R.H.
A-18916—REAR SEAT PAN
A-18898—REAR SEAT HEEL BOARD

A-156135—DOOR LOCK STRIKER PLATE
A-156030—FLOOR SIDE SILL ASSY. R.H.
A-157625—QUARTER HINGE PILLAR AND EXTENSION ASSY. R.H.

—Skeleton View of Town Sedan.

148

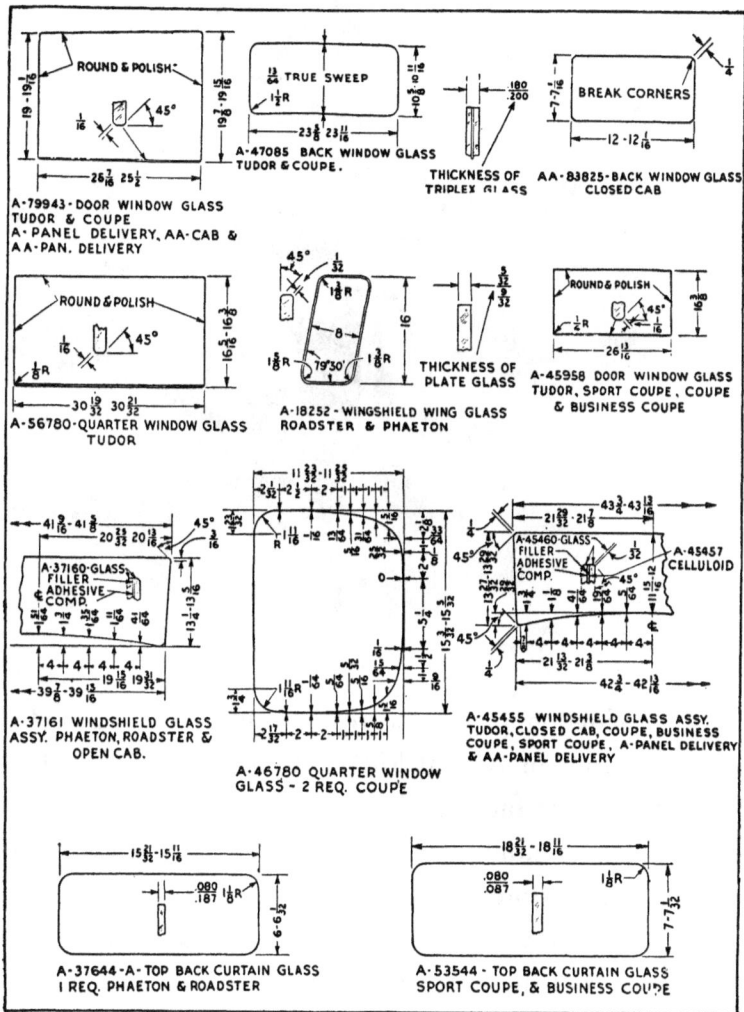

Ford Model A and AA Glass Chart Showing Sizes

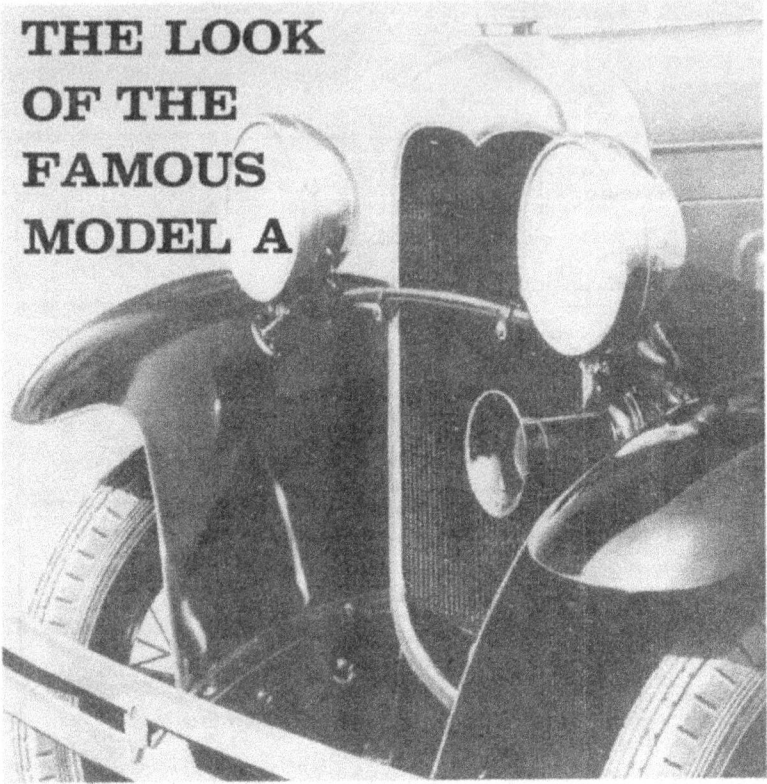

THE LOOK OF THE FAMOUS MODEL A

On the following pages the restorer will find a selection of photographs of typical Model A cars, 1928-1931. They are original Ford factory photographs (or renderings) and the negative file numbers are given following the car model number. The restorer who wants to study an illustration in greater detail, or would like to have it for his collection, can obtain copies of these pictures from Ford Research and Information Department, Ford Motor Co. Dearborn, Michigan. For a complete showing of all Model A's, approximately 250 photos, charts and drawings, see Floyd Clymer's "Model A Album".

1928 TUDOR SEDAN   55-A   50346

The Model T type bumper bars without end bolts indicate this to be a very early Tudor Sedan. Head lamps have one filament for driving (21 CP) and one for parking (3 CP); there is no "low" beam.

1928 STANDARD ROADSTER   40-A (Sport)   52187

A sporty accessory for MODEL A was the flying "Quail" ornament on the radiator cap. Note the two rumble seat steps, the larger brake drums, and the larger nickeled hub caps.

1928 STANDARD PHAETON   35-A   50359

Earliest of the MODEL A Fords, this Phaeton lacks outside door handles and separate parking brake system. Brake drums have chamfered edge and the small nickeled hub caps have a hexagonal bead.

1928 SPECIAL COUPE   49-A   51877

Body colors for 1928 were Arabian Sand, Niagara Blue, Gunmetal Blue, and Dawn Gray; fenders, wheels and running gear were Black. All bright work was nickel plated.

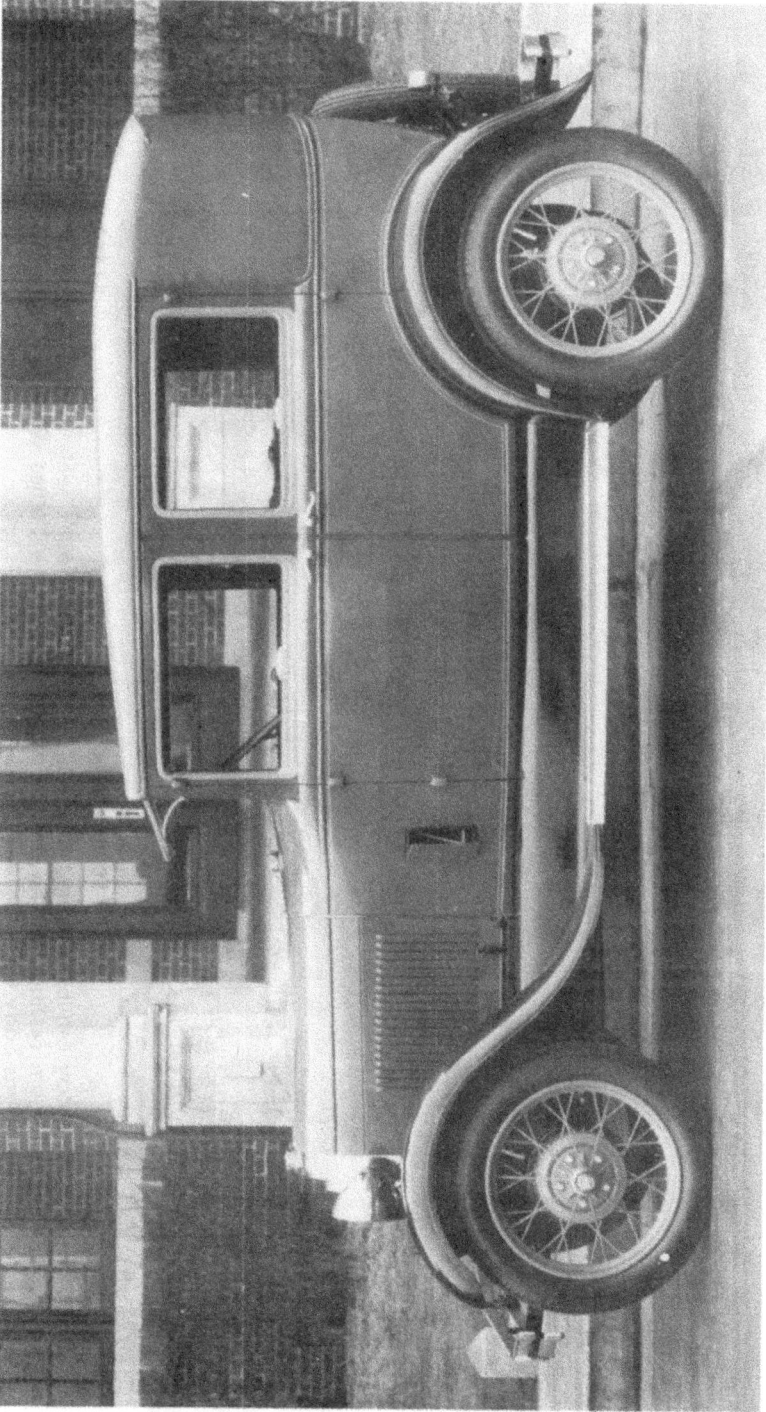

1928 FORDOR SEDAN   60-A   50831

The small, chamfered brake drums and the green celluloid sun visor mark this an early 1928 Ford. Note the open cowl ventilator on left side only.

1928 TOWN CAR 140-A 51423

A canopy can be snapped in place over the chauffeur's seat; behind is a large aluminum casting framing the rear windshield. These bodies were first supplied by Briggs Body Co.

1928 OPEN CAB (Roadster Pick-Up) 78-A 51174

Henry Ford insisted that the hub caps be attached so that their "Ford" script will appear upright when the wheel is mounted on the spare carrier with valve stem at the top.

1929 SPORT COUPE   50-A   53357

This closed Coupe looks like a convertible but isn't; the passengers in the rumble seat are the real "sports"!

1929 STANDARD ROADSTER   40-A   52383

This Balsam Green Roadster can be made weather-tight by snapping in place the two side curtains now stowed away under the seat.

1929 STANDARD PHAETON   35-A   52825

All Phaetons had black semi-gloss, grained tops; upholstering was unpleated artificial leather with a dark blue two-tone grained finish

1929 FODOR SEDAN 60-A 52703

This Briggs-built Fordor has a leather back with a Seal Brown fabric top. The Ford "Flying Quail" ornaments the radiator cap.

1929 CONVERTIBLE CABRIOLET   68-A   53573

First of the Ford Convertible Cabriolets, this speedy car has all the dash of the roadster and the comfort of the coupe. Upholstering is in artificial leather.

1929 TAXI CAB   135-A

Taxi Cabs were made only in late 1928 and 1929; the body is the only four door Ford to have the characteristic "coupe pillar" and the exposed cowl fuel tank. A "jump" seat is provided for a fourth passenger; baggage space is provided in front beside the driver.

163

1929 FORDOR SEDAN   60-B   58466

Another variation of the Fordor Sedan body by Briggs is this one having a leather back and black top. The little cylindrical tail lamp is an early 1929 carry-over from 1928.

1930 STANDARD PHAETON   35-B   54343

One man can easily put this top up in a few minutes — slightly longer in the rain. Side curtains are carried in a special compartment under the rear floorboard. "Standard" indicates the four-door Phaeton.

1930 DE LUXE PHAETON   180-A   55443-2

Even with top up, the two-door Phaeton is as streamlined as the cloche hat. Regular equipment includes a folding trunk rack on the back and the spare wheel mounted on the left side.

1930 DE LUXE ROADSTER   40-B   55693

A perfect car for the "lilt o' the links or the sweep o' the avenue" is this sport roadster. The Ford flying Quail symbolizes its quick get-away in all traffic.

1930 STANDARD COUPE   45-B   54854

Like all new Fords, the "Standard" Coupe is finished in a variety of beautiful colors. Its alert speed, quick acceleration, and ease of control give it a distinct advantage in crowded traffic.

1930 TUDOR SEDAN   55-B   54705

A conservative car of good appearance and quiet simplicity, the Tudor Sedan comfortably seats five passengers. Double-acting hydraulic shock absorbers make this an unusually easy riding car.

1930 CONVERTIBLE CABRIOLET   68-B   54838

Here is a smart all-weather car combining the airy freedom of the roadster with the snug comfort of the coupe — if you don't forget to put the top up. Who's for a seat full of snow?

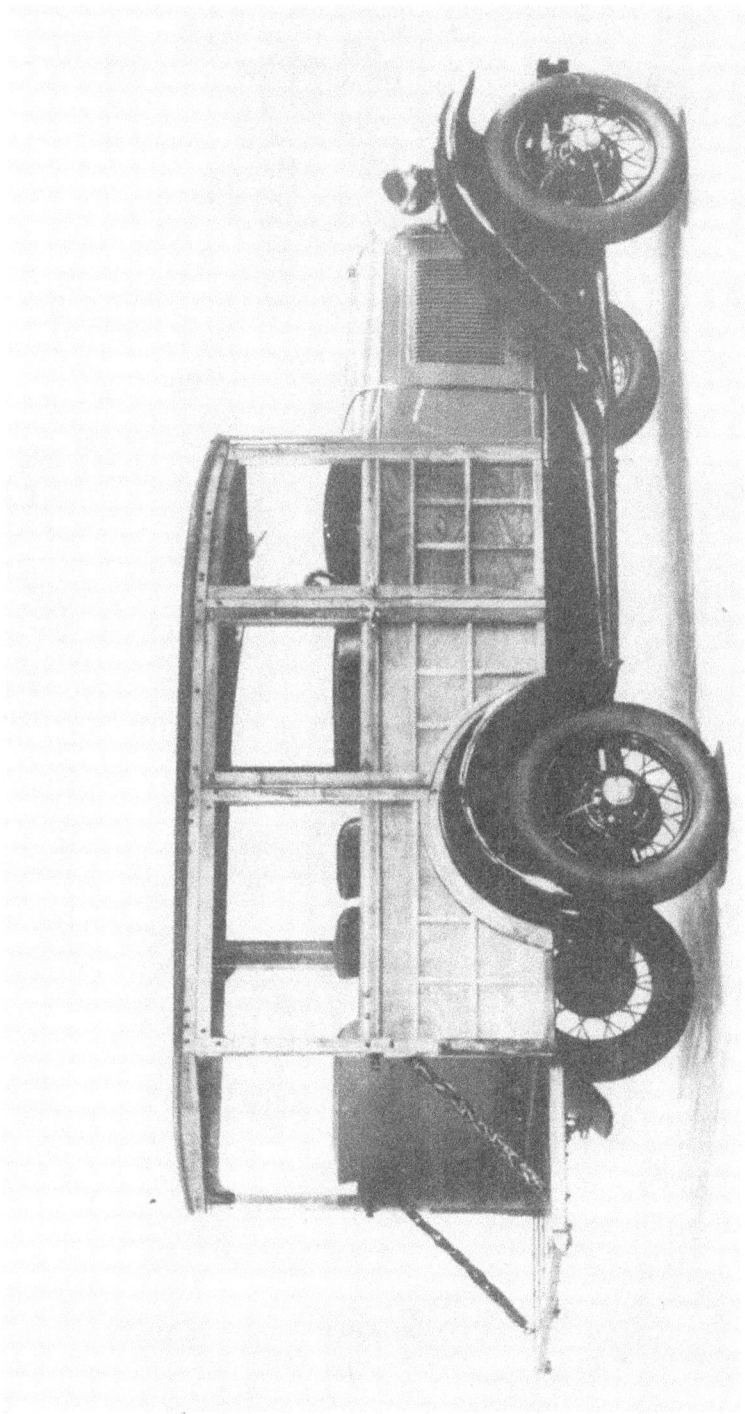

1930 STATION WAGON   150-B   54464

Ideal for week day hauling or week end calling, the Station Wagon has a large cargo space with tailgate down and rear seat removed. The maple body is sturdy enough for the heaviest of loads.

1930 TOWN SEDAN   155-C or -D   54351-1

The cowl lamps and the disappearing armrest in the center of the rear seat are some of the luxurious items differentiating this Town Sedan from the similar appearing Fordor Sedan.

78   **1930 TOWN SEDAN   155-C or -D   55447-10**

Model 155-C is built by Briggs; 155-D by Murray for Ford. Note the open space at top of rear seat cushion into which the center arm rest can be folded. Rear floor is recessed to give more passenger space.

73   1930 CONVERTIBLE CABRIOLET   68-B   55444-2

The "landau" top folds down in a jiffy to provide a car appropriate for the sport. The flying Quail mascot on the radiator symbolized the "get up and go" of the Ford Cabriolet.

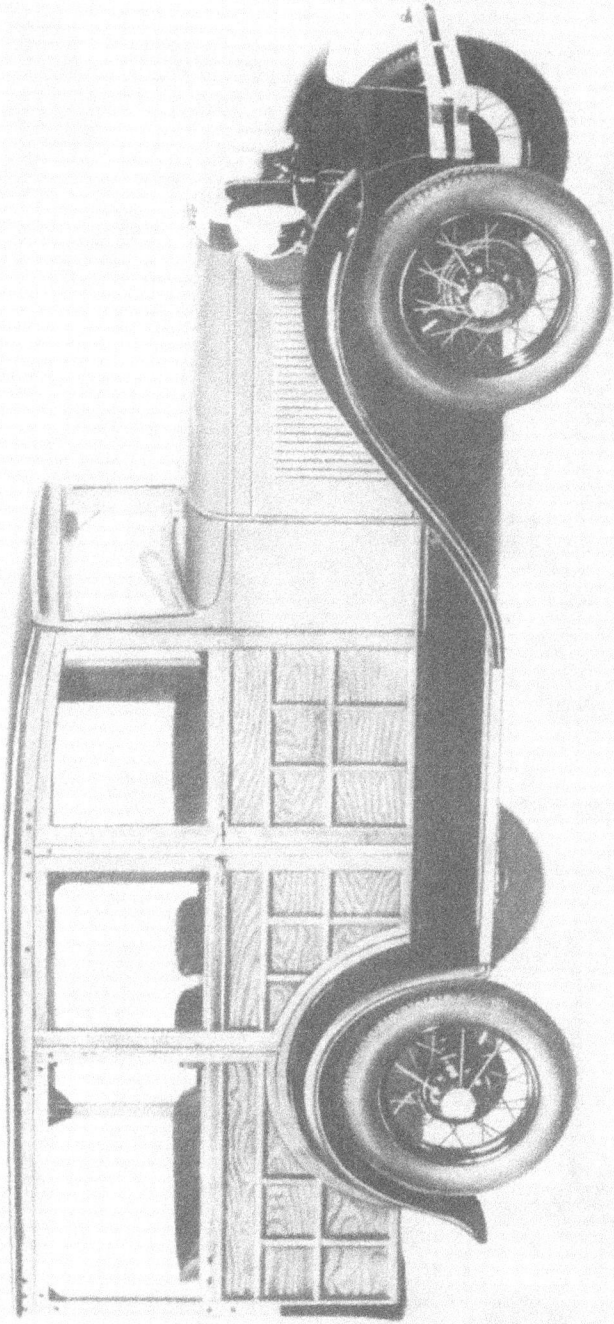

74 1930 STATION WAGON 150-B 55272

Ford pioneered assembly line production of station wagons; this is an excellent estate car for all-weather use. The maple wood body is built by Briggs or Murray.

58   **1930 SPORT PHAETON (Special)**   **55048**

This special body designed by Edsel Ford quite markedly resembles the contemporary British sports car, Bentley. This would indeed have been the "poor man's sports car" if the American public could have bought it!

59 **1930 SPORT PHAETON (Special)** **54046**

This double-cowl phaeton has a torpedo-like body; the top edges gracefully roll inward to form the double cockpits and the "bustle" rear rounds out the car in the best of 1930 sports car styling.

1931 STANDARD PHAETON   35-B   56360-3

This four door touring car is upholstered in imitation leather and carries the spare wheel on the back. Note the one-piece splash apron above running board.

1931 DE LUXE ROADSTER   **40-B**   **56058-12**

This sport roadster is finished in choice of Thorne Brown, Washington Blue, Stone Gray, Brewster Green, or Black. Cowl lamps and rumble seat are regular equipment.

1931 CABRIOLET (Convertible)   68-C   56197-2

A car for every season, this Cabriolet resembles the Sport Coupe when top is raised. Upholstering is either Tan Bedford cord or genuine crushed grain leather; rumble seat is two-tone cross grain artificial leather.

1931 TOWN SEDAN   155-D   56033

The visorless, slanting windshield gives a more regal look to the Town Sedan. A recessed rear floor affords more headroom for passengers.

1931 VICTORIA COUPE   190-A   56215-2

Especially graceful in appearance, this Victoria has a lower windshield and roof than any of the other closed Fords because of the extra deep floor recess. Separate front seats tip forward to provide access to the rear through the wide door.

182

1931 DE LUXE FORDOR SEDAN   170-B   56236-1

Though less expensive than the regal Town Sedan, this De Luxe Fordor has a very aristocratic appearance with its "blind" rear quarter panel which affords the passengers seclusion if not a view.

1931 CONVERTIBLE SEDAN   400-A   56379-5

This Convertible is the stylists' delight with its Copra Drab body, Chickle Drab reveals, Tan top, and Tacoma Cream wheels all set off by the Black fenders and splash aprons.

105　　**1931 CABRIOLET (Convertible)**　　**68-C**　　**56266-1**

This Bronson Yellow and Seal Brown two-tone Cabriolet with chromium plated windshield frame is the height of fashion; it looks sleek even with top up.

122　　1931 SPORT COUPE　　50-B　　55793

Clearly shown here are the dummy landau irons on the top and the one-piece splash apron characteristic of the 1931 Fords. Except for the military visor and the lack of cowl lamps, this car resembles the convertible Cabriolet.

128   1931 CONVERTIBLE SEDAN   400-A   56655-2

This is the newest Ford body creation — with windows up and top down, passengers enjoy all advantages of an open air car with none of its disadvantages.

92   **1931 DE LUXE PHAETON**   **180-A**   **56671-5**

A sport car for the entire family, this two-door Phaeton has such de luxe equipment as cowl lamps, genuine leather upholstering, side mirror, left side mounted spare wheel, and folding trunk rack on rear.

123   1931 STATION WAGON   150-B   56228

The multi-purpose Station Wagon is growing in popularity. The tan engine hood and cowl harmonize admirably with the varnished maple wood body. Briggs and Murray continue to supply these bodies to Ford for factory assembling.

NOTES

NOTES

NOTES

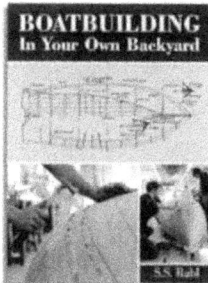

www.ingramcontent.com/pod-product-compliance
Lightning Source LLC
Chambersburg PA
CBHW070330090426
42733CB00012B/2424